ENERGY SCIENCE, ENGINEERING AND TECHNOLOGY

CHALLENGES FACING THE NUCLEAR POWER INDUSTRY

ENERGY SCIENCE, ENGINEERING AND TECHNOLOGY

Additional books in this series can be found on Nova's website under the Series tab.

Additional e-books in this series can be found on Nova's website under the e-book tab.

ENERGY SCIENCE, ENGINEERING AND TECHNOLOGY

CHALLENGES FACING THE NUCLEAR POWER INDUSTRY

RENEE OLSON
EDITOR

nova publishers
New York

Library of Congress Cataloging-in-Publication Data

ISBN: 978-1-63483-618-0

Published by Nova Science Publishers, Inc. † New York

CONTENTS

PREFACE

The nuclear industry is in a state of flux. Proclamations of a nuclear renaissance have given way to predictions of the industry's ultimate decline, and now, in the face of new carbon emissions regulation, renewed optimism for the future. Yet, nuclear plants are currently at-risk due to electricity market design issues and competition with low priced natural gas. This book attempts to provide clarity to the potential outlook for the nuclear industry by examining the factors that influence it. It includes an analysis of recent trends affecting nuclear power generation while focusing on the specific potential impact of the EPA's proposed Clean Power Program on nuclear power.

In: Challenges Facing the Nuclear Power Industry ISBN: 978-1-63483-618-0
Editor: Renee Olson

Chapter 1

ISSUE PAPER: ELECTRICITY SYSTEM ADEQUACY - CHALLENGES FACING THE NUCLEAR POWER INDUSTRY*

C. Elise Logan, John Brewer and Robert Wallace

ACRONYMS AND ABBREVIATIONS

AEO 2014	Annual Energy Outlook 2014	LSE	Load Serving Entity
CC	Combined cycle	MW	Megawatt
CO2	Carbon dioxide	NARUC	National Association of Regulatory
COL	Combined Operating License		Utility Commissioners
DOE	Department of Energy	NERC	North American Electric Reliability
EIA	Energy Information Administration		Corporation
EISPC	Eastern Interconnection States'	NETL	National Energy Technology
	Planning Council		Laboratory
EPA	Environmental Protection Agency	NRC	Nuclear Regulatory Commission

* This is an edited, reformatted and augmented version of a report, DOE/NETL-DOE/NETL-2015/1702, prepared under contract by Energy Sector Planning and Analysis (ESPA) for the United States Department of Energy (DOE), National Energy Technology Laboratory (NETL), issued November 6, 2014.

ESPA	Energy Sector Planning and Analysis	O&M	Operating and maintenance
GW	Gigawatt	TVA	Tennessee Valley Authority
kW	Kilowatt	U.S.	United States

1. INTRODUCTION

The nuclear industry is in a state of flux, because the last several years have seen earlier proclamations of a nuclear renaissance give way to predictions of the industry's ultimate decline, and now, in the face of new carbon emissions regulation, renewed optimism for the future. After coasting steadily through the 1990s, with profitable plant operations, few retirements, and rising capacity through uprates, the nuclear industry began to look forward to a possible revival in plant builds. In 2007, the Nuclear Regulatory Commission (NRC) received the first applications for new nuclear reactors that had been submitted since 1978. [1] Five nuclear reactors are currently under construction[1] and more are in advanced stages of development, which is a significant increase in activity since the last nuclear plants went into operation during the late 1980s and early 1990s.

In the midst of this activity, the nuclear power industry has begun to experience new challenges. As natural gas prices have dropped, bringing down the price of electricity as well, the economics of nuclear plants have begun to look less competitive. Operators of existing nuclear power plants are now experiencing lower cost margins due, in part, to the increased competition from gas-fired generation. [2] Further, estimates of the cost of constructing new nuclear plants have increased from earlier predictions, and new regulations, such as those adopted by the NRC after the 2011 Fukushima accident, are driving up costs for existing reactors. Instead of a nuclear renaissance, some analysts are now predicting a decline for the nuclear industry, arguing that nuclear power is too expensive to compete in energy markets. [3] Seemingly bearing out this prediction, operators of six nuclear reactors, representing nearly 5,000 MW of generating capacity, have either shut down or announced retirement plans since 2012 [4], and still others have cancelled planned uprates. [5]

Those predictions have not taken into account considerations of carbon emissions regulation. In April 2014, the Environmental Protection Agency (EPA) announced its proposed "*Clean Power Plan*" [2]. This plan, which could change significantly before it becomes a final rule, would reduce carbon emissions from the electric power industry, and it includes nuclear power as an option for meeting state emissions goals. [6] This caused analysts to predict that nuclear power may still have a future. [4] However, there is still a great deal of uncertainty surrounding compliance with the *Clean Power Plan*, which allows states flexibility in choosing how they will comply. In addition, as discussed further in this report, the EPA itself does not appear to expect nuclear power to play a major role in compliance with the *Clean Power Plan*. The EPA's model predicts that state goals will be met by decreased use of coal-fired generation paired with increased gas-fired generation, renewables, and efficiency. [7]

This report attempts to provide clarity to the potential outlook for the nuclear industry by examining the factors that influence it. It includes an analysis of recent trends affecting nuclear power generation while focusing on the specific potential impact of the EPA's proposed Clean Power Program on nuclear power.

2. Current Trends in Nuclear Generation

There are currently 100 nuclear reactor units operating in the United States (U.S.), totaling over 100,000 MW of operating capacity. Nearly all of these reactors were built prior to 1990. Aside from a few retirements, and even fewer units coming online, the number of operating units remained fairly static throughout the 1990s. During this period, additional operating capacity was mainly achieved through uprates, which increase the power output of an existing unit through an equipment upgrade. [8] Since then, however, there has been increased interest in constructing new nuclear reactors. From 2007 to 2009, the NRC has received applications to construct new nuclear facilities, called Combined Operating Licenses (COL), for 28 new units. [9] Of those applications, the NRC has approved 4 new units, 12 units are still under review, and applications for the remaining 12 units have been either postponed or withdrawn by the applicants. Additionally, the Tennessee Valley Authority (TVA) resumed construction of the Watts Bar Unit 2 reactor in 2007, which had been suspended for two decades and is now expected to go into service in 2015.

The NRC's process for reviewing units was stalled for two years, following a June 8, 2012 ruling by the U.S. Court of Appeals for the DC Circuit that found that the NRC's Waste Confidence Rule did not satisfy its obligations per the National Environmental Protection Act (NEPA).

Exhibit 2-1. Nuclear power reactors in the U.S. [9, 10]

Status	Number of Units	MW Operating Capacity
Operating	100	100,715
Under Construction	5	5,538
COL Under Review	12	15,587
Postponed/Withdrawn COL Application	12	14,495
Retired/Planned Retirement 2013-2019	6	4,836

The court found that in assessing the environmental impacts of a potential nuclear plant, the NRC should consider the impact of failing to secure permanent disposal for spent nuclear fuel, as well as the impact of potential spent fuel pool leaks and fires. In response to the ruling, the NRC chose to suspend licensing decisions that rely on the Waste Confidence Rule until a revised rule could be developed. [11] During this time, the NRC continued to work on licensing activities, but did not issue any licenses for new nuclear facilities or for relicensing of existing plants. On August 26, 2014, the NRC voted to approve a new rule on waste analysis and to resume making decisions on licenses. [12, 13]

In contrast to the spike in COL applications to the NRC, 6 units have either retired or announced their near-term retirement since 2013. Two of the units, San Onofre Units 2 and 3 in San Diego, California, were retired early once it was determined that it was not economic to restart them. [14] The units had been shut down after a leak was discovered in one steam generator, and the cost to repair the leak and ensure the reliability of the plant's other steam generators was too high to justify keeping San Onofre in operation. [15] Similarly, the Crystal River reactor in Citrus County, Florida, was retired in 2013 rather than repaired, after it experienced problems with a steam generator replacement. [16] The Kewaunee plant in Wisconsin also retired in 2013, although it was operating well and had just extended its operating license for another 20 years. Driven by cheap natural gas and an influx of available wind energy, power prices in the Midwest had declined to the point where the Kewaunee plant was operating at a loss and unable to compete in the energy markets. [17]

The Vermont Yankee and Oyster Creek nuclear reactors have announced their retirements, for 2014 and 2019, respectively, both citing high operating costs and low energy prices as driving factors in the decision to retire. In both cases, the economic issues facing the plants were made more difficult because of state level opposition. In the case of Oyster Creek, the New Jersey Department of Environmental Protection would have required the plant to install cooling towers, estimated to cost $700 million, if Oyster Creek had not chosen to retire early. [18] The state legislature in Vermont passed legislation that would have forced Vermont Yankee to shut down in 2012. Although Entergy Corporation, Vermont Yankee's owner, successfully sued to keep the plant open, the legal battles contributed to the difficulties it faced in continuing operations. [19, 20] As discussed further below, more units may be at-risk of retirement if certain economic trends continue.

As shown in Exhibit 2-2, existing nuclear plants are scattered across the Eastern Interconnection, with only a few plants located in the West. Much of the recent activity in nuclear development has been concentrated in the southeastern U.S., as seen in Exhibit 2-3. The five nuclear reactor units currently being constructed are located in Georgia (Vogtle Units 3 and 4), South Carolina (V.C. Summer Units 1 and 2), and Tennessee (Watts Bar Unit 2). Among the COL applications under review by the NRC, which have not been suspended, most are located in East Coast states.

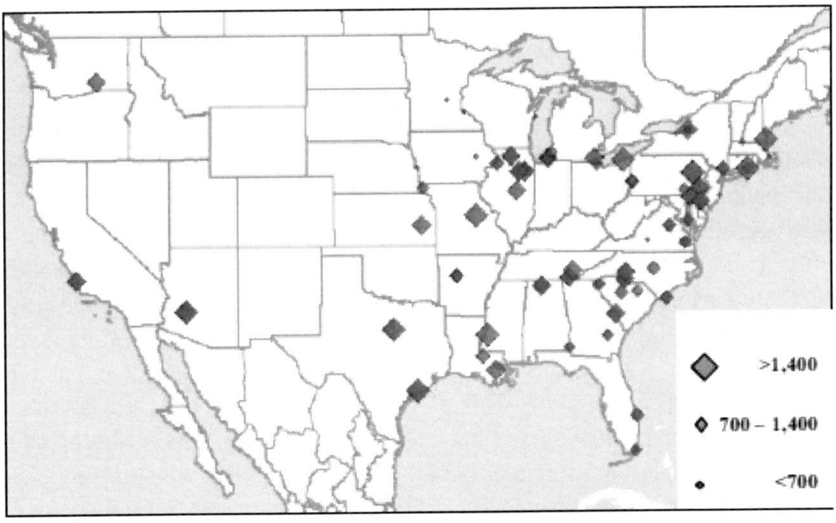

Source: NETL.

Exhibit 2-2. Nuclear power reactors currently operating in the U.S. (sized by MW operating capacity).

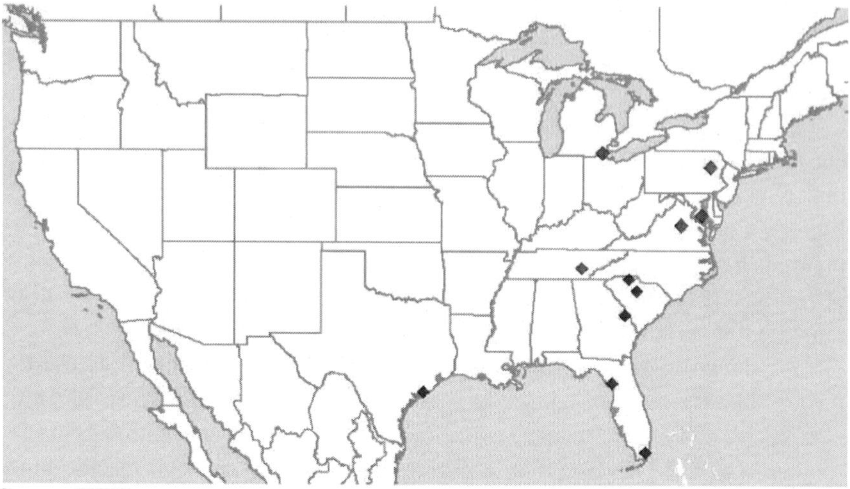

Source: NETL.

Exhibit 2-3. Planned nuclear units under construction and/or NRC review (9).

In its Annual Energy Outlook 2014 (AEO 2014), the Energy Information Administration (EIA) models several different possible outlooks for nuclear power. Under the Reference case scenario projected in the AEO 2014, EIA assumes that known economical, technological, and demographical trends will continue. [21] For nuclear power, it assumes that the units currently under construction will come online on schedule, and that the existing units that have announced their retirements will retire as planned. The total overnight costs for constructing a new advanced nuclear reactor are assumed to be $5,501/kW, which is in line with the estimates for the four new units currently under construction. [21] The Vogtle reactors are expected to cost $6,800/kW, and the VC Summer units are expected to cost $4,772/kW. [22] In comparison, EIA estimates that the overnight cost for constructing an advanced gas-fired combined cycle (CC) plant is $1,021/kW, while a scrubbed coal plant is $2,925/kW. [21]

EIA estimates that the fixed operating and maintenance (O&M) costs for an advanced nuclear reactor are $93.28/kW/year. [21] These estimated costs are considerably higher than EIA's estimates for other new generating units. For example, an advanced CC plant's estimated fixed O&M costs are $15.37/kW/year and a scrubbed coal plant's costs are $31.18/kW/year. Even the estimated costs for an integrated coal-gasification CC plant with carbon sequestration are lower than for a nuclear plant, at $72.84/kW/year.[3] [21] The estimated variable O&M costs for an advanced nuclear reactor, however, are only $2.14/kW/year. This is lower than the cost for fossil fuel generation, which EIA estimates is $4.47/kW for scrubbed coal and $3.26/kW for advanced CC. However, it exceeds the $0/kW variable O&M costs for wind, solar, and geothermal renewables.

The AEO includes projections for three possible variations for nuclear power, apart from the Reference case – Accelerated Nuclear Retirements, Low Nuclear, and High Nuclear.

- The Accelerated Nuclear Retirements case assumes that all nuclear plants are limited to a 60-year lifespan; uprates have been limited to the 0.7 GW that have been reported to EIA, and there are no new nuclear power units beyond those currently under construction. In addition to the 4.8 GW of announced retirements accounted for in the Reference case, the Accelerated Nuclear Retirements case also includes a decrease of 5.7 GW in capacity to account for plants at risk of early retirement. It assumes that nonfuel operating costs will increase by 3% per year after 2013.
- The Low Nuclear case combines the assumptions of the Accelerated Nuclear Retirements case with those in the High Oil and Gas Resource and the No Sunset cases.[5] The assumptions from High Oil and Gas Resource case forecast more favorable conditions for natural gas-fired and renewable generation, depressing the economic outlook for nuclear power.
- The High Nuclear case assumes that all nuclear plants, except for the 4.8 GW of announced retirements, are life-extended beyond 60 years. It also assumes a total of 6.0 GW of uprates at existing plants. It includes new plants currently under construction as well as those that have completed COL applications with NRC, for a total of 12.6 GW of planned capacity additions. [21]

In the Reference case, overall energy production from nuclear power remains flat through 2040, as shown in Exhibit 2-4. Under the High Nuclear case, there is a net increase of 20.8 GW of nuclear capacity from 2013 to 2040. The Advanced Nuclear Retirements and Low Nuclear cases see dramatic declines in capacity of 38.5 GW and 73.8 GW, respectively. Under the Low Nuclear case, retirements rapidly accelerate after 2030 so that by 2040, only 25.2 GW of capacity remain. The Low Nuclear Case makes clear the negative impact that abundant natural gas could have on nuclear capacity.

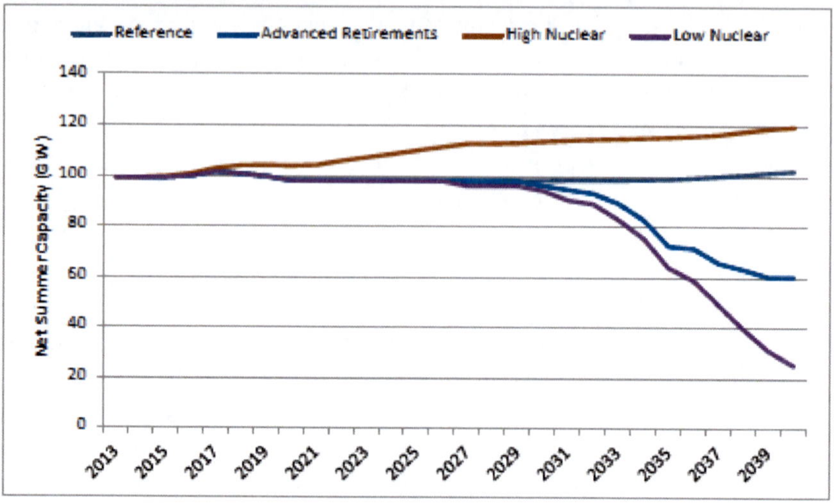

Exhibit 2-4. AEO 2014 projections of nuclear capacity.

As the different cases modeled in the AEO 2014 show, the outlook for the nuclear power industry in the U.S. varies dramatically depending on the set of assumptions being used. Although the AEO 2014 offers several different scenarios, it does not provide a full explanation for the drivers behind the different assumptions in each case. To understand which of these outlooks is more accurate, a better context is needed for the economic and regulatory circumstances facing the nuclear industry.

Recent studies on the state of the nuclear power industry in the U.S. have identified several key challenges facing the nuclear industry, naming economic conditions as the biggest of those challenges. In 2013, the Eastern Interconnection States' Planning Council (EISPC) and the National Association of Regulatory Utility Commissioners (NARUC) published the "Assessment of the Nuclear Power Industry" (Nuclear Assessment). [8] In addition to serving as a primer on the history of the nuclear industry and offering an overview of future nuclear technology, the Nuclear Assessment identifies the issues currently faced by the industry. The Vermont Law School has put out two reports on the state of the nuclear industry – one in 2009 and one in 2013. Both Vermont Law reports argue that the nuclear industry is facing an imminent decline. [3, 23]

The Nuclear Assessment finds that the nuclear industry faces a significant timing challenge. Reviewing the outlook for replacing anticipated coal plant retirements, the Nuclear Assessment found that natural gas power compared much more favorably than nuclear power, even when factoring in natural gas infrastructure constraints. According to the Nuclear Assessment, a natural gas plant can be constructed in two-to-three years following the receipt of permits. A new pipeline serving that plant can be sited, approved by regulatory authorities, and constructed within three years. In contrast, the COL application process for a new nuclear reactor takes four to six years, and then subsequent construction is five to six years. [8] Considering that a significant number of coal plant retirements are projected to take place over the next five years, nuclear plants not already under construction could not be constructed quickly enough to replace a retiring coal plant.

Noting the recent retirements of plants facing economic difficulties, the Nuclear Assessment finds that the significant additions of new nuclear baseload generation envisioned are unlikely to be built. According to the estimates provided in the Nuclear Assessment, the capital costs of building a nuclear reactor would have to be reduced by half in order to be competitive with a CC natural gas plant under the current economic climate. It further claims that the "nuclear renaissance" may be a phenomenon confined to the southeastern U.S. [8] The regulatory environment in the southeastern U.S. is more favorable to nuclear power than many other regions because it does not have deregulated energy markets. Deregulated energy markets place all of the risk for construction delays, project financing, and cost estimates on the plant investor. This is riskier for companies building a nuclear plant than for a regulatory regime, because a regulatory regime allows the investor to place some of the risk on customers. [9] All of the nuclear plants currently under construction are in states with traditional regulatory structures, and most of the plants that have retired or announced retirements were operating in deregulated energy markets.

The 2009 Vermont Law report finds that, compared to preliminary industry estimates of the costs to build a new nuclear plant that were issued in the early 2000s, more recent estimates can be as much as four times higher. [3] The cost overruns that are experienced at the reactors currently under construction confirms that the real cost of new nuclear reactors tends to be higher than early estimates predicted. For example, the costs to complete the Watts Bar Unit 2 reactor were estimated at $2.5 billion; the budget for completion is now $4 billion and has been delayed by three years. [22] The 2009 Vermont Law report concluded that nuclear power is not competitive with other forms of electric generation. According to the report, even under carbon emission reduction policies, there are multiple other generation options, particularly wind and natural gas, which are lower cost than new nuclear reactors. [3]

The 2013 Vermont Law report focuses on the economics of existing nuclear units, identifying nuclear reactors that are at risk of early retirement. Most of the nuclear generation fleet has been operating for decades. As the AEO 2014's Low Nuclear case illustrates, the majority of the nuclear fleet will reach the end of their 60-year lifespans between 2030 and 2040. [21] Aging plants are often more costly to maintain, and thus, they are at greater risk of retiring earlier due to maintenance problems. As discussed above, three of the units that retired in 2013 did so because repairs were too expensive.

According to the 2013 Vermont Law report, 38 nuclear power plants are at risk of early retirement. [3] Two of those plants – Oyster Creek and Vermont Yankee – have already announced that they will retire soon. The 2013 Vermont Law report evaluated the plants based on a number of risk factors, including costs, age, and size of the facility, whether or not it is operating in a deregulated market, its history of reliability and long-term outages, and whether it would require a retrofit to comply with new Fukushima-related regulations. The report finds that risk factors related to costs, reliability, and repairs increase with a plant's age.

Regulation also has a significant effect on nuclear plants, particularly as post-Fukushima regulations go into effect at the same time nuclear plants are experiencing financial difficulties. Since the accident at the Japanese Fukushima Dai-ichi plant in 2011, the NRC has reviewed the circumstances of the event to evaluate the need for additional regulation to prevent or mitigate such an accident at U.S. nuclear facilities. As a result of this review, the NRC is enacting several new requirements, including the construction of containment venting systems at plants with designs similar to the Fukushima plant, the addition of spent fuel pool instrumentation, and the implementation of other mitigation strategies. A number of other possible regulations are still under consideration. [24]

The industry has recently begun to push back against the regulatory burdens that they feel the NRC has placed on them, and the NRC has implemented a pilot program in six nuclear plants aimed at addressing some of these concerns. Specifically, the pilot program attempts to prioritize the regulations that the nuclear operators must comply with, and allows operators some flexibility in how and when they reach compliance. [25] In another example of industry pushback, the Department of Energy (DOE) has suspended its collection of the nuclear waste disposal fee after the industry successfully challenged it in court. [26] This does not, however, solve the spent fuel storage problem. The Nuclear Assessment points out that long-term storage of spent fuel may be the most complex problem faced by the nuclear industry, which is a problem that cannot be solved under the current regulatory structure. The nuclear industry is dependent on a political solution to the problem of long-term storage, something that has been debated without resolution since the first nuclear power reactors began operation in 1956.

3. Potential Impact of Greenhouse Gas Regulation

The EPA's *Clean Power Plan* could potentially improve the outlook for the nuclear power industry. Nuclear power generation is specifically included in the *Clean Power Plan* as a possible way to reduce carbon emissions. The proposed *Clean Power Plan* designates four building blocks for how states can reduce their electric power sector carbon emissions. They are:

1) Increase efficiency of existing coal-fired plants
2) Increase utilization of natural gas
3) Increase utilization of renewables and nuclear
4) Increase the amount of energy efficiency [6]

States do not need to use all of these building blocks to comply with the *Clean Power Plan.* Instead, these four building blocks can be mixed and matched in different ways – giving states maximum flexibility in deciding how best to comply with the rule. According to the EPA, policies that encourage development of new nuclear capacity and discourage premature retirement of nuclear capacity could be useful elements of a carbon reduction strategy. [6]

3.1. Clean Power Plan

The *Clean Power Plan*, as proposed, requires each state to design a plan to meet state-specific rate-based goals for lowering carbon emissions from the power sector. These state-specific goals must be met by 2030. The state goals established by the *Clean Power Plan* were not developed from a uniform baseline (i.e., requiring states to reduce their carbon emissions by a certain percentage based on emissions from a specific year), but rather from the

EPA's assessment of the amount of emissions that can be reduced at existing fossil-fired electric generating units, at a reasonable cost, through application of the four building blocks. [6] The EPA developed a base case that modeled each state's expected emissions for the year 2030 (absent any new regulations), which it used in developing its assessment of the amount of emissions each state could achieve. Thus, the emissions goals vary by state.

The EPA's methodology for developing the state goals relies on its assessment of emissions from all existing electric generating units within each state. With respect to nuclear power, the EPA considered all currently operating nuclear plants as existing plants, except Vermont Yankee, which has announced plans to retire in 2014. "At-risk" plants (those that might be shut down) were also included in the analysis as were the five units currently under construction as existing units. [27] This means that when the Vogtle, V.C. Summer, and Watts Bar reactors come online, they will be considered part of their respective state's base case generation mix, and will not count toward meeting emissions goals. If they were to fail to come online, or if any "at-risk" plants were to retire, their state's emissions would fall further from the emissions goals set by the EPA. A state that experiences the shutdown of a large nuclear plant would be faced with having to make up the loss of non-carbon emitting generation in addition to the actions it would already need to take in order to meet its emission reduction goals. This will likely place pressure on states to ensure that nuclear plants within their boundaries remain in operation.

In its regulatory impact analysis of the *Clean Power Plan*, the EPA modeled the impact of compliance on the generation mix. Compared to the base case, the EPA predicts that nuclear capacity will remain unchanged under the *Clean Power Plan*. In both the base case and the compliance case, EPA predicted that nuclear power generation will fall from 103 GW in 2020 to 101 GWh in 2030. [7] Notwithstanding the inclusion of nuclear power as one of the building blocks for compliance, the EPA itself does not appear to expect nuclear power to play a major role in compliance with the *Clean Power Plan*. The EPA's model instead predicts that state goals will be met by increased retirements of coal-fired generation paired with increased gas-fired generation, renewables, and efficiency.

In its discussion of nuclear power's emissions reduction potential in the proposed rule, the EPA stated that it will accept comments on whether it is appropriate to include the five nuclear units under construction in a state's goals. The EPA acknowledged that reflecting completion of these units in developing a state's goals has a significant impact on the calculated goals for the states in which these units are located. It thus appears that this portion of the *Clean Power Plan* is still an open question. The EPA also stated that it is seeking comments on how it should factor at-risk nuclear capacity and planned nuclear construction into the calculation of state goals. [6]

3.2. Implications for Nuclear Power

To date, nuclear power has not been included by most states in their attempts to reduce carbon emissions. Of the 37 states that have adopted some form of renewable portfolio standard or goal, only Indiana and Ohio include nuclear power as part of their clean energy policies. [28, 29] Additionally, several states have policies expressly meant to restrict the development of nuclear reactors. [8] Exhibit 3-1 shows which states have adopted policies to either encourage or discourage the development of nuclear power generation. Policies that are considered a disincentive include, but are not limited to, restrictions on nuclear waste, requiring voter or legislature approval before a nuclear plant can be constructed, and Minnesota's ban on any development of new nuclear reactors.[5] Policies that are considered incentives include, among others, Kansas and Alabama's tax incentives for nuclear reactors and Indiana and Ohio's categorization of nuclear power as clean energy for the purpose of meeting clean energy policies. As demonstrated by the number of states with disincentives for nuclear development, as well as the lack of consideration of nuclear power as a source of clean energy, using nuclear power as a means to meet the *Clean Power Plan's* goals would constitute a shift in policy for many states.

The *Clean Power Plan* requires an average of about 30 percent in carbon emission reductions compared to the EPA's estimated 2030 base case for each state, although some state goals require as little as 10 or 12 percent and some require as much as 62 percent in reductions. For most states, constructing a new nuclear reactor alone would not be sufficient to meet their emissions goals. This does not mean, however, that states will not choose to use nuclear power as one of the building blocks in their state plans. There are several states with pro-nuclear policies that would be more likely to embrace nuclear expansion in their plans for compliance. For those states where plans for new nuclear reactors are already underway, continuing with their development could prove to be a cost-effective option. Those states with limited renewable resources may also need to turn to nuclear power generation in order to meet their goals.

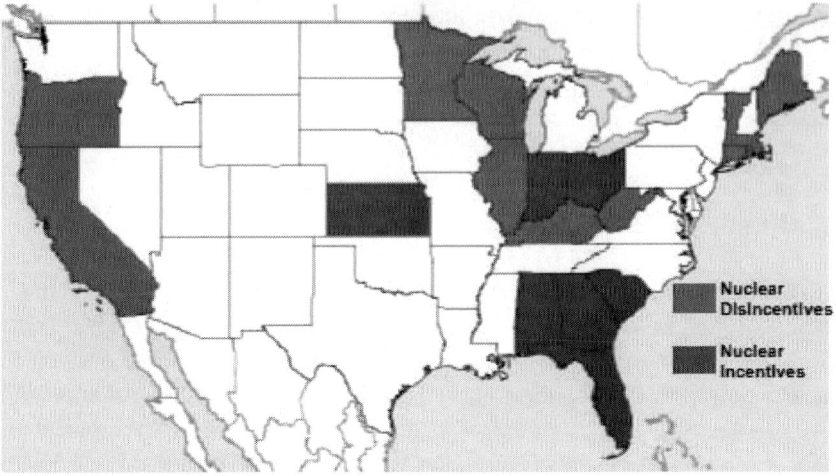

Source: NETL.

Exhibit 3-1. State policies on nuclear power generation [8].

The analysis in this report indicates that the states most likely to include nuclear power are located in the southeastern U.S. As shown in Exhibit 3-2, these states already have policies and regulatory regimes in place that support nuclear development as well as nuclear projects that are either under construction or being developed.

Exhibit 3-2. Factors encouraging nuclear development in southeast

State	Traditional regulatory structure	Pro-nuclear policies	Nuclear projects under construction or development
Alabama	Yes	Yes	Yes
Florida	Yes	Yes	Yes
Georgia	Yes	Yes	Yes
South Carolina	Yes	Yes	Yes
Tennessee	Yes	No	Yes

CONCLUSION

Despite earlier predictions of a nuclear renaissance, the economics facing both new and existing nuclear power reactors remain challenging. Nuclear reactors in deregulated markets are experiencing declining cost margins as low gas prices and increasing amounts of renewables push energy prices lower. Existing reactors are also experiencing rising costs as maintenance costs at aging plants rise and new regulations impose additional burdens on the nuclear industry. The outlook for nuclear power in states with a traditional regulatory structure remains more positive. Most of the nuclear reactors under construction or development are located in these regulated states.

It appears that even under the proposed *Clean Power Plan*, large amounts of nuclear plants are unlikely to be built. Most states will be able to more cost-effectively reach their emissions goals by relying on other means, such as replacing coal plants with natural gas-fired generation and providing greater incentives for renewable sources of electricity. The most significant impact of the *Clean Power Plan* may be in providing states with the incentive to prevent aging nuclear plants from retiring in order to prevent the loss of a major non-carbon-emitting source of power. Rather than a nuclear renaissance characterized by the construction of new nuclear reactors and the continued relicensing of older ones, the outlook for the nuclear industry is likely to be more modest. The *Clean Power Plan* may slow the number of retirements that would otherwise have occurred, but it does not provide states with significant incentives to add more nuclear capacity.

As the proposed rule is finalized, however, the *Clean Power Plan* could change significantly. As such, continued monitoring of its potential impacts on the nuclear power industry is important. Additionally, the expected shift from baseload coal and nuclear sources of electricity to increasing amounts of gas-fired and renewable generation could have significant impacts on the reliability of the bulk power system. The *Clean Power Plan's* potential impact on grid reliability could cause cost increases that have not been accounted for in this report and which could change the economic calculus for some nuclear reactors on the margin.

REFERENCES

[1] Wellock, J. Samuel Walker and Thomas R. *A Short History of Nuclear Regulation, 1946- 2009.* Washington, DC : Nuclear Regulatory Commission (NRC), 2010.

[2] The Economist. Fracked off: Thanks to cheap natural gas, America's nuclear renaissance is on hold. *The Economist.* [Online] June 1, 2013. [Cited: July 16, 2014.] http://www.economist.com/news/united-

states/21578690-thanks-cheap-natural-gas-americasnuclear-renaissance-hold-fracked.

[3] Cooper, Mark. *The Economics of Nuclear Reactors: Renaissance or Relapse?* South Royalton, VT : Vermont Law School, 2009.

[4] Hiar, Corbin. S&P: Climate rules could produce 'some increased interest' in new nuclear plants . *SNL.* [Online] June 6, 2014. [Cited: July 15, 2014.] http://www.snl.com/InteractiveX/article.aspx?Id=28315145&KPLT=2.

[5] Brandyk, Matthew. NPPD cancels plans to boost capacity at Cooper nuclear plant. *SNL.* [Online] August 12, 2013. [Cited: July 14, 2014.] http://www.snl.com/interactivex/article.aspx?id=19101185&KPLT=6.

[6] Environmental Protection Agency (EPA). Carbon Pollution Emission Guidelines for Existing Stationary Sources: Electric Utility Generating Units. *Federal Register.* [Online] June 18, 2014. [Cited: July 11, 2014.] https://www.federalregister.gov/articles/2014/06/18/2014-13726/carbon-pollution-emission-guidelines-for-existing-stationary-sources-electric-utilitygenerating.

[7] —. *Regulatory Impact Analysis for the Proposed Carbon Pollution Guidelines for Existing Power Plants and Emission Standards for Modified and Reconstructed Power Plants.* Washington, DC : EPA, 2014.

[8] Navigant Consulting, Inc. *Assessment of the Nuclear Power Industry: Final Report.* Washington : Eastern Interconnection States' Planning Council and National Association of Regulatory Utility Commissioners, 2013.

[9] Nuclear Regulatory Commission (NRC). Combined License Applications for New Reactors. *NRC.* [Online] July 1, 2014. [Cited: July 21, 2014.] http://www.nrc.gov/reactors/newreactors/col.html.

[10] SNL Energy. Custom Report Builder. *SNL Energy.* [Online] [Cited: July 22, 2014.] http://www.snl.com/SNLWebPlatform/Content/News/News.aspx.

[11] Nuclear Regulatory Commission (NRC). Waste Confidence. *NRC.* [Online] September 4, 2014. [Cited: September 5, 2014.] http://www.nrc.gov/waste/spent-fuel-storage/wcd.html.

[12] Bandyk, Matthew. NRC lifts nuke licensing moratorium. *SNL.* [Online] August 26, 2014. [Cited: September 5, 2014.] http://www.snl.com/interactivex/article.aspx?id=28991089&KPLT=6.

[13] —. Update: NRC approves rule on spent fuel storage. *SNL.* [Online] August 26, 2014. [Cited: September 5, 2014.] http://www.snl.com/interactivex/article.aspx?id=29002054&KPLT=6.

[14] Testa, Dan. Update: As delays, uncertainty grew, SoCalEd's decision to retire San Onofre took shape. *SNL.* [Online] June 7, 2013. [Cited: August 12, 2014.] http://www.snl.com/interactivex/article.aspx?id=17941384&KPLT=6.

[15] Khan, Zia Ullah. SoCalEd to take hefty Q2 charge for San Onofre nuke retirement. *SNL.* [Online] June 2007, 2013. [Cited: August 12, 2014.] http://www.snl.com/InteractiveX/article.aspx?ID=17940568.

[16] Bandyk, Matthew. Update: Duke to retire Crystal River nuke, reviews plans for replacement power. *SNL.* [Online] February 5, 2013. [Cited: August 12, 2014.] http://www.snl.com/InteractiveX/article.aspx?ID=16916418.

[17] Engblom, Andrew. Update: Low power prices from shale gas boom force closure of Dominion's Kewaunee nuke. *SNL.* [Online] October 22, 2012. [Cited: August 12, 2014.] http://www.snl.com/interactivex/article.aspx?id=16038231&KPLT=6.

[18] Barber, Wayne. Exelon's Oyster Creek to close within decade, avoid $700M in cooling tower costs . *SNL.* [Online] December 9, 2010. [Cited: August 12, 2014.] http://www.snl.com/interactivex/article.aspx?id=12081248&KPLT=6.

[19] Qureshi, Nazia. Entergy to retire, decommission Vermont Yankee nuclear plant. *SNL.* [Online] August 27, 2013. [Cited: August 12, 2014.] http://www.snl.com/interactivex/article.aspx?id=24717220&KPLT=6.

[20] Bandyk, Matthew. Update: Appeals Court: State overstepped authority on Vermont Yankee nuclear plant. *SNL.* [Online] August 13, 2013. [Cited: August 15, 2014.] http://www.snl.com/interactivex/article.aspx?id=24490486&KPLT=6.

[21] U.S. Energy Information Administration (EIA). *Assumptions to the Annual Energy Outlook 2014.* Washington, DC : EIA, 2014.

[22] Bandyk, Mathew. End in sight for TVA at one of the longest nuclear projects ever. *SNL.* [Online] August 14, 2014. [Cited: August 14, 2014.] http://www.snl.com/Interactivex/article.aspx?ID=28847750.

[23] Cooper, Mark. *Renaissance in Reverse: Competition Pushes Aging U.S. Nuclear Reactors to the Brink of Economic Abandonment.* South Royalton : Vermont Law School, 2013.

[24] Nuclear Regulatory Commission (NRC). Japan Lessons Learned. *NRC.* [Online] April 24, 2014. [Cited: August 13, 2014.] http://www.nrc.gov/reactors/operating/ops-experience/japandashboard.html.

[25] Brandyk, Matthew. NRC regulations under industry scrutiny in era of closing nuclear reactors. *SNL.* [Online] July 7, 2014. [Cited: July 11, 2014.] http://www.snl.com/interactivex/article.aspx?id=28525510&KPLT=6.

[26] Wolff, Eric. Department of Energy zeroes out nuclear waste fee. *SNL.* [Online] May 15, 2014. [Cited: July 21, 2014.] http://www.snl.com/InteractiveX/article.aspx?ID=28134145.

[27] Environmental Protection Agency (EPA). EPA's Power Sector Modeling Platform v.5.13. *EPA.* [Online] November 27, 2013. [Cited: August 13, 2014.] http://www.epa.gov/powersectormodeling/BaseCasev513.html.

[28] U.S. Energy Information Administration (EIA). Most States have Renewable Portfolio Standards. *Energy Information Administration.* [Online] February 3, 2012. [Cited: August 4, 2014.] http://www.eia.gov/todayinenergy/detail.cfm?id=4850&src=email.

[29] North Carolina State University. *DSIRE - Database of State Incentives for Renewables and Efficiency.* [Online] [Cited: August 4, 2014.] http://www.dsireusa.org/incentives/index.cfm?EE=0&RE=1&SPV=0&ST=0&searchtype=RPS &sh=1.

End Notes

[1] Four of the units are V.C. Summer Units 1 and 2 in South Carolina, and Vogtle Units 3 and 4 in Georgia. A fifth unit, Watts Bar Unit 2 in Tennessee, is also under construction and expected to go into service as early as 2015. The Watts Bar Unit 2 was originally granted a construction permit in 1973; due to multiple deficiencies and complications, construction was suspended in 1986. In 2007, Tennessee Valley Authority (TVA) informed the NRC of its intention to resume construction.

[2] Environmental Protection Agency (EPA) Standards of Performance for Greenhouse Gas Emissions from Existing Sources: Electric Utility Generating Units. Federal Register June 18, 2014. Also referred to by industry as the proposed "Existing Source Greenhouse Gas" regulations.

[3] See Table 8.2 of "Assumptions to the Annual Energy Outlook 2014" for a complete listing of estimated costs.

[4] The High Oil and Gas Resource case assumes that recovery per shale gas, tight gas, and tight oil well is 50 percent higher and well spacing is 50 percent lower than in the Reference case. It also adds new tight oil resources and includes 50 percent higher undiscovered resources than in the Reference case. The No Sunset case assumes the extension of all existing tax

credits and policies that contain sunset provisions, except those requiring additional funding or extensive regulatory analysis.

[5] Minnesota does have three existing nuclear reactors still operating within the state – Monticello and Prairie Island Units 1 and 2.

This report was prepared by Energy Sector Planning and Analysis (ESPA) for the United States Department of Energy (DOE), National Energy Technology Laboratory (NETL). This work was completed under DOE NETL Contract Number DE-FE0004001. This work was performed under ESPA Task 150.09.01.

The authors wish to acknowledge the excellent guidance, contributions, and cooperation of the NETL staff, particularly: *Maria A. Hanley, NETL Technical Monitor, Thomas J. Tarka, P.E. NETL TCOR*

In: Challenges Facing the Nuclear Power Industry ISBN: 978-1-63483-618-0
Editor: Renee Olson

Chapter 2

NUCLEAR ENERGY: OVERVIEW OF CONGRESSIONAL ISSUES*

Mark Holt

SUMMARY

The policy debate over the role of nuclear power in the nation's energy mix is rooted in the technology's fundamental characteristics. Nuclear reactors can produce potentially vast amounts of energy with relatively low consumption of natural resources and emissions of greenhouse gases and other pollutants. However, facilities that produce nuclear fuel for civilian power reactors can also produce materials for nuclear weapons. The process of nuclear fission (splitting of atomic nuclei) to generate power also results in the production of radioactive material that must be contained and can remain hazardous for thousands of years. How to manage the weapons proliferation and safety risks of nuclear power, or whether the benefits of nuclear power are worth those risks, are issues that have long been debated in Congress.

The 100 licensed nuclear power reactors at 62 sites in the United States generate about 20% of the nation's electricity. Five new reactors are currently under construction. About a dozen more are planned, but whether they move forward will depend largely on their economic

* This is an edited, reformatted and augmented version of a Congressional Research Service publication R42853 prepared for Members and Committees of Congress dated March 14, 2014.

competitiveness with natural gas and coal plants. Throughout the world, 435 reactors are currently in service, and 71 more are under construction.

The March 2011 disaster at the Fukushima Dai-ichi nuclear power plant in Japan increased attention to nuclear safety throughout the world. The U.S. Nuclear Regulatory Commission (NRC), which issues and enforces nuclear safety requirements, established a task force to identify lessons from Fukushima applicable to U.S. reactors. The task force's report led to NRC's first Fukushima-related regulatory requirements on March 12, 2012. Several other countries, such as Germany and Japan, eliminated or reduced their planned future reliance on nuclear power after the accident.

Highly radioactive spent nuclear fuel that is regularly removed from nuclear power plants is currently stored at plant sites in the United States. Plans for a permanent underground repository at Yucca Mountain, NV, were abandoned by the Obama Administration, although that decision is being challenged in court. The Obama Administration appointed the Blue Ribbon Commission on America's Nuclear Future to recommend an alternative nuclear waste policy. In response to the Commission's recommendations, the Department of Energy (DOE) issued a new waste strategy in January 2013 that calls for the selection of new candidate sites for nuclear waste storage and disposal facilities through a "consent-based" process and for a surface storage pilot facility to open by 2021.

The level of security that must be provided at nuclear power plants has been a high-profile issue since the 9/11 terrorist attacks on the United States in 2001. Since those attacks, NRC issued a series of orders and regulations that substantially increased nuclear plant security requirements, although industry critics contend that those measures are still insufficient.

Encouraging exports of U.S. civilian nuclear products, services, and technology while making sure they are not used for foreign nuclear weapons programs has long been a fundamental goal of U.S. nuclear energy policy. Recent proposals to build nuclear power plants in several countries in the less developed world, including the Middle East, have prompted concerns that international controls may prove inadequate.

Synthesis of Key Issues

The long-running policy debate over the future of nuclear energy is rooted in the technology's inherent characteristics. Initially developed for its unprecedented destructive power during World War II, nuclear energy seemed to hold equal promise after the war as a way of providing limitless energy to all mankind. International diplomacy has focused ever since on finding institutional mechanisms for spreading the perceived benefits of nuclear energy throughout the world while preventing the technology from being used for the proliferation of nuclear weapons. Much of this international effort is focused on key nuclear fuel cycle facilities—plants for enriching uranium in the fissile isotope U-235 and for separating plutonium from irradiated nuclear fuel. Such plants can be used to produce civilian nuclear reactor fuel as well as fissile material for nuclear warheads.

Yet even the use of nuclear power solely for peaceful energy production has proven intrinsically controversial. The harnessing of nuclear fission in a reactor creates highly radioactive materials that must be kept from overheating and escaping from the reactor building, as occurred during the disasters at Fukushima and Chernobyl. Spent nuclear fuel that is regularly removed from reactors during refueling must be isolated from the environment for up to a million years. Potential technologies to reduce nuclear waste through recycling usually involve separating plutonium that could be used for nuclear weapons and would still leave substantial amounts of radioactive waste to be stored and disposed of. Long-term storage and disposal sites for nuclear waste have proven difficult to develop throughout the world, as illustrated by the Obama Administration's cancellation of the proposed U.S. waste repository at Yucca Mountain, NV.

The March 2011 disaster at Japan's Fukushima Dai-ichi nuclear power plant, which forced the evacuation of areas as far as 30 miles away, has slowed nuclear power expansion plans around the world, particularly in Japan and Western Europe. However, dozens of new reactors are still being planned and built in China, India, Eastern Europe, and elsewhere.[1] In these areas, nuclear power's initial promise of generating large amounts of electricity without the need for often-imported fossil fuels, along with the more recent desire to reduce greenhouse gas emissions, remains a compelling motivation.

With 100 licensed reactors, the United States has the largest nuclear power industry in the world. But U.S. nuclear power growth has been largely stagnant for the past two decades, as natural gas has captured most of the market for new electric generating capacity.[2] Congress enacted incentives for new nuclear plants in the Energy Policy Act of 2005 (P.L. 109-58), including production tax credits, loan guarantees, and insurance against regulatory delays. Those incentives, combined with rising natural gas prices and concerns about federal restrictions on carbon dioxide emissions, prompted industry plans by late 2009 for up to 30 new nuclear power reactors in the United States.[3] However, falling natural gas prices and the defeat of greenhouse gas legislation in the 111th Congress have put many of those projects on hold. Currently, four new reactors, in Georgia and South Carolina, are under construction, and an older reactor on had been suspended for two decades is now being completed in Tennessee. A variety of incentives to renew the growth of nuclear power have been proposed, including a plan by President Obama to include nuclear power, along with natural gas and advanced coal technologies, in a federal mandate for the production of "clean energy."

Existing U.S. nuclear power plants are facing difficult competition from natural gas and renewable energy. Four U.S. reactors were permanently closed in 2013, and the shutdown of a fifth unit was announced for late 2014. Three of those units closed because of the need for expensive repairs, while the other two were operating well but could not compete in their local wholesale electricity markets. All five units had substantial time remaining on their initial 40-year operating licenses or had received or applied for 20-year license extensions from the Nuclear Regulatory Commission (NRC). The shutdowns prompted widespread discussion about the future of other aging U.S. reactors. The extent to which the growth of nuclear power should be encouraged in the United States and around the world will continue to be a major component of the U.S. energy policy debate. Questions for Congress will include the implementation of policies to encourage or discourage nuclear power, post-Fukushima safety standards, development of new nuclear power and fuel cycle technologies, and nuclear waste management strategies.

BASIC FACTS AND STATISTICS

The 100 licensed nuclear power reactors at 62 sites in the United States generate about 20% of the nation's electricity. The oldest of today's operating reactors were licensed in 1969, and the most recent was in 1996. The reactors were initially licensed to operate for 40 years, but 80% have received or applied for 20-year license renewals by NRC. Under the current mixture of 40- and 60-year licenses, 33 reactors would have to shut down by 2030 and the rest by 2049.[4]

Whether new reactors will be constructed to replace the existing fleet or even to expand nuclear power's market share will depend largely on costs. The cost of building and operating a new nuclear power plant in the United States is generally estimated to be significantly higher than natural gas combined-cycle plants (which use both combustion and steam turbines) and somewhat above conventional coal-fired plants. For example, the Energy Information Administration (EIA) estimates that electricity generation from a nuclear power plant coming on line in 2018 would cost 10.8 cents per kilowatt-hour (kwh), while advanced combined-cycle gas would cost 6.6 cents/kwh, and conventional coal would cost 10.0 cents/kwh. EIA estimates that onshore wind would cost 8.7 cents/kwh, offshore wind 22.2 cents/kwh, and solar photovoltaic 14.4 cents/kwh.[5] Such estimates depend on a wide range of variables, however, such as future fuel costs and environmental regulations. Targeted tax credits and other incentives for specific technologies, which are not included in the EIA estimates, would also affect nuclear power's economic competitiveness.

As noted above, the United States currently has five reactors under construction. They are scheduled to begin operating in 2015 through 2018.[6] Licenses to build and operate 12 additional reactors are currently pending at NRC, although some of their review schedules are uncertain.[7] If those additional U.S. reactors are licensed and built, they could begin coming on line in the early 2020s.

Throughout the world, 435 reactors are currently in service or operable, and 71 more are under construction. France is the most heavily nuclear-reliant country in the world, with 58 reactors generating 75% of the country's electricity in 2012. Thirty countries in 2012 generated at least some of their electricity from nuclear power. After the Fukushima accident, Germany, which had previously generated about 30% of its electricity with nuclear power, closed eight of the country's 17 power reactors and decided to shut the remainder by 2022.[8] Japan, which had also generated about 30% of its electricity with nuclear power and had planned to raise that level to 50%, is reconsidering its energy policy. None of Japan's 48 operable reactors is currently running. Safety improvements in response to the tsunami are currently being implemented, and 17 reactors are undergoing regulatory reviews for possible restart. It is not clear how many of the operable reactors will ultimately seek restart approval.

MAJOR NUCLEAR ENERGY ISSUES

Safety

The Fukushima Dai-ichi disaster, triggered by a huge earthquake and tsunami, greatly increased concerns about safety in the nuclear policy debate. The accident clearly demonstrated the potential consequences of a total loss of power (or "station blackout") at today's commercial nuclear plants. Even when a reactor shuts down, as the Fukushima plant did after the initial earthquake, residual radioactivity in the reactor core continues to generate heat that must be removed, typically by electrically driven or controlled cooling systems. When the tsunami knocked out power at three of the Fukushima reactors, the buildup of heat and pressure became so great that it melted the reactors' nuclear fuel and exceeded the limits of their containment structures. Cooling was also lost in Fukushima's spent fuel storage pools, causing concern that they could overheat, although later examination indicated that they did not.

Safety requirements for nuclear power plants are established and enforced in the United States by NRC, an independent regulatory commission. NRC safety regulations address the effects of external events such as earthquakes and floods, equipment failure such as breaks in coolant pipes, and other problems that could lead to radioactive releases into the environment. Critics of nuclear power contend that NRC is often reluctant to impose necessary safety requirements that would be costly or disruptive to the nuclear industry. However, the industry has frequently contended that costly safety proposals are unnecessary and would not significantly increase large existing safety margins.

Recent Events

Following the Fukushima disaster, NRC established a task force to identify lessons applicable to U.S. reactors and recommend safety improvements. The task force's report led to NRC's first Fukushima-related regulatory requirements, on March 12, 2012. NRC ordered all reactors to develop strategies to maintain cooling and containment integrity during external events, such as floods and earthquakes, that were more severe than anticipated by the plants' designs ("beyond design basis"). In addition, NRC required that U.S. reactors of similar design to the Fukushima reactors have "reliable hardened vents" to remove excess pressure from their primary containments, and that better instrumentation be installed to monitor the condition of spent fuel pools during accidents.[9] The NRC commissioners on March 19, 2013, required NRC staff to study whether to require the newly mandated containment vents to include filters or other means to reduce the release of radioactive material if the vents have to be used. The idea of requiring filters had drawn praise from nuclear critics but opposition from the industry on cost grounds.[10]

Selected Congressional Action

Nuclear Regulatory Commission Reorganization Plan Codification and Complements Act (H.R. 3132, Terry, S. 1519, Vitter)

Specifies functions and authorities of the Chairman and Commissioners of NRC. Specifies that any commissioner may request a vote on whether a particular issue should be reserved for the Chairman or handled by the full Commission. House bill introduced September 18, 2013; referred to Committee on Energy and Commerce. Senate bill introduced September 18, 2013; referred to Committee on Environment and Public Works. Hearing held by House Committee on Energy and Commerce, Subcommittees on Energy and Power and Environment and the Economy, December 12, 2013.

Oversight Hearing: NRC's Implementation of the Fukushima Near-Term Task Force Recommendations and Other Actions to Enhance and Maintain Nuclear Safety

Joint hearing before the Senate Committee on Environment and Public Works and Subcommittee on Clean Air and Nuclear Safety. January 30, 2014. Witnesses: NRC Chairman Allison M. Macfarlane and NRC commissioners.

Additional References

Nuclear Safety: Countries' Regulatory Bodies Have Made Changes in Response to the Fukushima Daiichi Accident, Report to the Chairman, Subcommittee on Transportation and Infrastructure, Committee on Environment and Public Works, U.S. Senate, Government Accountability Office, GAO-14-109, March 2014, http://www.gao.gov/products/GAO-14-109.

Recommendations for Enhancing Reactor Safety in the 21st Century, Nuclear Regulatory Commission, Near-Term Task Force Review of Insights from the Fukushima Dai-ichi Accident, July 12, 2011, http://pbadupws.nrc.gov/docs/ML1118/ML111861807.pdf.

State-of-the-Art Reactor Consequence Analyses (SOARCA) Report: Draft Report for Comment, Nuclear Regulatory Commission, NUREG-1935, January 2012, http://www.nrc.gov/reading-rm/ doc-collections/nuregs/staff/sr1935.

Radioactive Waste

Highly radioactive spent nuclear fuel must regularly be removed from operating reactors and stored in adjacent pools of water. After several years of cooling, the spent fuel can be placed in dry casks for storage elsewhere on the plant site. When existing U.S. reactors were built, spent fuel had been expected to be taken away for reprocessing (separation of plutonium and uranium to make new fuel) or permanent disposal. However, reprocessing has not become commercialized in the United States, for economic and nonproliferation reasons, and central waste storage and disposal facilities have proven difficult to site. As a result, the vast majority of U.S. commercial spent fuel remains at the nuclear plants where it was generated—totaling 71,775 metric tons in 2013 and rising at the rate of about 2,000 metric tons per year.[11]

Recent Events

The Nuclear Waste Policy Act (P.L. 97-425, NWPA), as amended in 1987, named Yucca Mountain, NV, as the nation's sole candidate site for a permanent high-level nuclear waste repository. However, the Obama Administration decided to halt the Yucca Mountain project and appointed the Blue Ribbon Commission on America's Nuclear Future to recommend an alternative policy. The Commission issued its final report in January 2012, and the Department of Energy (DOE) responded in January 2013 with a new waste strategy that calls for a "consent-based" process to select nuclear waste storage and disposal sites and for a surface storage pilot facility to open by 2021.[12] A federal appeals court on August 13, 2013, ordered NRC to continue the Yucca Mountain licensing process with previously appropriated funds.[13] NRC responded November 18, 2013, by directing the agency's staff to complete the Yucca Mountain safety evaluation report, a key document that would provide the staff's conclusions about whether the proposed repository could be licensed.

Selected Congressional Action

No More Excuses Energy Act of 2013 (H.R. 2081, Thornberry)

Includes provisions to prohibit NRC from considering nuclear waste storage when licensing new nuclear facilities, and to establish a tax credit for obtaining nuclear component manufacturing certification. Introduced May 21, 2013; referred to multiple committees.

Dry Cask Storage Act (H.R. 3354, Engel)

Requires spent fuel at nuclear power plants to be moved from spent fuel pools to dry casks after it has sufficiently cooled. Costs of the fuel transfers would be offset by a reduction in nuclear waste fees owed to the federal government. Introduced October 28, 2013; referred to Committee on Energy and Commerce.

Nuclear Waste Administration Act of 2013 (S. 1240, Wyden)

Establishes an independent Nuclear Waste Administration to develop nuclear waste storage and disposal facilities. Siting of such facilities would require the consent of the affected state, local, and tribal governments. The Nuclear Waste Administration could spend nuclear waste fees collected after the bill's enactment without the need for further appropriation. Fee collection would halt after 2025 if a waste facility had not been opened. Introduced June 27, 2013; referred to Committee on Energy and Natural Resources. Committee hearing held July 30, 2013.

Oversight Hearing: Nuclear Waste Programs and Strategies

Oversight hearing by the House Appropriations Committee Energy and Water Development and Related Agencies Subcommittee, Thursday, April 11, 2013. Lead witnesses: Peter B. Lyons, Assistant Secretary for Nuclear Energy, Department of Energy, and Michael Weber, Deputy Executive Director, Nuclear Regulatory Commission.

Additional References

Report to the Secretary of Energy, Blue Ribbon Commission on America's Nuclear Future, January 2012, http://cybercemetery.unt.edu/archive/brc/20120620211605/http:/brc.gov.

Managing Spent Nuclear Fuel: Strategy Alternatives and Policy Implications, RAND Corporation, 2010, http://www.rand.org /pubs/monographs/MG970.html.

Federal Support and Incentives

Congress has long debated the role that nuclear power should play in meeting national energy and environmental goals. Nuclear power supporters generally point to the technology as crucial for providing a secure, domestic source of energy with low greenhouse gas and other emissions. Opponents generally counter that safety and proliferation risks, nuclear waste hazards, and high costs outweigh those benefits. The debate over nuclear power's role often focuses on the level of federal support that should be provided to encourage the construction of new nuclear plants, through such mechanisms as loan guarantees, tax credits, clean energy mandates, and liability limits under the Price-Anderson Act. Because of the relatively high cost of new nuclear reactors, especially compared with natural gas plants, the level of federal support is expected to be a key determinant of the future growth or decline of nuclear power in the United States. Federal funding for nuclear energy research and development, along with related infrastructure and security, is debated annually in Congress as part of the Energy and Water Development appropriations bill. DOE nuclear energy funding totals $888.4 million for FY2014, while for FY2015 the President is requesting a reduction to $863.4 million.

Recent Events

One nuclear power project, consisting of two new reactors at the Vogtle plant in Georgia, received a conditional commitment from DOE for an $8.33 billion loan guarantee in February 2010, as authorized by Section 1703 of the Energy Policy Act of 2005 (P.L. 109-58). Energy Secretary Ernest Moniz announced the issuance of $6.5 billion in loan guarantees on February 19, 2014, to two of the three utility partners in the project, Georgia Power and Oglethorpe Power. An additional $1.8 billion loan guarantee for another partner, Municipal Electric Authority of Georgia, is still pending. No other planned nuclear plants have received conditional commitments for DOE loan guarantees.

Selected Congressional Action

Energy Freedom and Economic Prosperity Act (H.R. 259, Pompeo)

Terminates nuclear energy production tax credit, among other provisions. Introduced January 15, 2013; referred to Committee on Ways and Means.

Additional References

Maintaining U.S. Leadership in Global Nuclear Energy Markets, Bipartisan Policy Center, September 2012, http://bipartisanpolicy.org /sites/default/files/Nuclear%20Report.PDF.

World Nuclear Industry Status Report 2013, Mycle Scheider Consulting, July 30, 2013, http://www.worldnuclearreport.org/-2013-.html.

Security and Emergency Response

The level of security that must be provided at nuclear power plants has been a high-profile issue since the 9/11 terrorist attacks on the United States in 2001. Since those attacks, NRC issued a series of orders and regulations that substantially increased nuclear plant security requirements, although industry critics contend that those measures are still insufficient. Key measures include an increase in the level of attacks that nuclear plant security forces must be able to repel, requirements for mitigating the effects of large fires and explosions, and a requirement that new reactors be capable of withstanding aircraft crashes without releasing radioactive material. NRC also modified its planning requirements for evacuations and other emergency responses after the 9/11 attacks, and the Fukushima disaster illustrated the importance of emergency response to radioactive releases from any cause.

Recent Events

NRC issued wide-ranging revisions to its emergency preparedness regulations on November 1, 2011, dealing with duties of emergency personnel and the inclusion of hostile actions in emergency planning drills.[14] In response to Fukushima, NRC staff recommended that nuclear emergency plans be required to address events affecting multiple reactors and prolonged station blackout. NRC told nuclear power plants on March 12, 2012, to provide specific information and analysis on those issues.[15]

Selected Congressional Action

Pandemic and All-Hazards Preparedness Reauthorization Act of 2013 (P.L. 113- 5, H.R. 307)

Reauthorizes public health security programs, including those related to nuclear and radiological hazards. Introduced January 18, 2013, by Representative Mike Rogers of Michigan; passed House January 22, 2013, by vote of 395-29. Passed Senate by unanimous consent with an amendment February 27, 2013. House suspended rules and agreed to Senate amendment 370-28 on March 4, 2013. President signed March 13, 2013.

Nuclear Disaster Preparedness Act (H.R. 1700, Engel)

Requires the President to issue guidance for federal response to nuclear disasters, covering specific topics listed in the bill. Introduced April 24, 2013; referred to Committee on Transportation and Infrastructure.

Additional References

Protecting Our Nation, Nuclear Regulatory Commission, NUREG/BR-0314, Rev. 3, October 2013, http://pbadupws.nrc.gov/docs/ML1327 /ML1327 0A213.pdf.

Nuclear Weapons Nonproliferation

Encouraging exports of U.S. civilian nuclear products, services, and technology while making sure they are not used for foreign nuclear weapons programs has long been a fundamental goal of U.S. nuclear energy policy. International controls and inspections are intended to ensure the peaceful use of civilian nuclear facilities and prevent the proliferation of nuclear weapons. However, recent proposals to build nuclear power plants in as many as 18 countries[16] that have not previously used nuclear energy, including several in the Middle East and elsewhere in the less developed world, have prompted concerns that international controls may prove inadequate. Numerous recommendations have been made in the United States and elsewhere to create new incentives for nations to forgo the development of uranium enrichment and spent nuclear fuel reprocessing facilities that could produce weapons materials as well as civilian nuclear fuel.

Recent Events

Iran is currently the prime example of the tension between peaceful and weapons uses of nuclear technology. Of particular concern is a growing Iranian uranium enrichment program, which Iran contends is solely for peaceful purposes but which the United States and other countries suspect is for producing weapons material. The U.N. Security Council has imposed sanctions and passed several resolutions calling on Iran to suspend its enrichment program and other sensitive nuclear activities. Nevertheless, Iran continues to advance its nuclear program.

Extension of the U.S.-South Korea nuclear cooperation agreement, running through March 19, 2014, has also been affected by nonproliferation issues. South Korea would like to include advance U.S. consent for spent fuel reprocessing and uranium enrichment, but the United States is concerned about the precedent that such an agreement might set and how it would affect other ongoing issues on the Korean peninsula. Legislation authorizing the President to extend the existing nuclear cooperation agreement by two years was signed into law on February 12, 2014 (P.L. 113-81), allowing more time to negotiate a new, long-term agreement.

Selected Congressional Action

Iran, North Korea, and Syria Nonproliferation Accountability Act of 2013 (H.R. 893, Ros-Lehtinen)

Imposes penalties for nuclear technology trade with Iran, North Korea, and Syria, restricts U.S. cooperation with countries aiding nuclear weapons proliferation in Iran, North Korea, and Syria, and establishes related sanctions. Introduced February 28, 2013; referred to multiple committees.

To Amend the Atomic Energy Act of 1954 to Require Congressional Approval of Agreements for Peaceful Nuclear Cooperation with Foreign Countries, and for Other Purposes (H.R. 3766, Ros-Lehtinen)

Requires congressional approval of U.S. peaceful nuclear cooperation agreements with countries that do not agree to forgo enrichment and reprocessing. Introduced December 12, 2013; referred to Committees on Foreign Affairs and Rules.

Hearing: Civilian Nuclear Cooperation Agreements

Hearing by the Senate Committee on Foreign Relations, January 30, 2014. Examined whether U.S. nuclear cooperation agreements should require cooperating countries to forgo uranium enrichment and reprocessing and the potential impact of a declining U.S. role in global nuclear trade. Lead witnesses: Thomas M. Countryman, Assistant Secretary of State, Bureau of International Security and Nonproliferation, and Daniel B. Poneman, Deputy Secretary of Energy.

Other References

Nuclear Nonproliferation: IAEA Has Made Progress in Implementing Critical Programs but Continues to Face Challenges, Government Accountability Office, GAO-13-139, May 16, 2013, http://www.gao. Gov / products/GAO-13-139.

U.S. Nuclear Cooperation as Nonproliferation: Reforms, or the Devil You Know? Nuclear Threat Initiative, November 27, 2012, http:// www . nti .org /analysis/articles/us-nuclear-cooperationnonproliferation-reforms-or-devil-you-know.

End Notes

[1] World Nuclear Association, "World Nuclear Power Reactors & Uranium Requirements," February 1, 2014, http://www.world-nuclear.org/info/ Facts-and-Figures/World-Nuclear-Power-Reactors-and-Uranium-Requirements.

[2] Energy Information Administration, "Most Electric Generating Capacity Additions in the Last Decade Were Natural Gas-Fired," July 5, 2011, http://www.eia.gov/todayinenergy/ detail.cfm?id=2070.

[3] Nuclear Regulatory Commission, "Expected New Nuclear Power Plant Applications," updated September 28, 2009. Available from the author

[4] Nuclear Regulatory Commission, *Information Digest, 2032-2014*, NUREG-1350, Vol. 25, October 2013, Section 3: Nuclear Reactors, http://www.nrc.gov/reading-rm/doc-collections/nuregs/staff/sr1350.

[5] Energy Information Administration, "Levelized Cost of New Generation Resources in the Annual Energy Outlook 2013," January 2013, http://www.eia.gov/forecasts/aeo/pdf/ electricity_generation.pdf. Levelized costs include capital costs averaged over the life of the plant, plus fuel and maintenance costs.

[6] Tennessee Valley Authority, "Watts Bar Unit 2 Project Construction Update," October26, 2012, http://www.tva.com/ power/nuclear/wattsbar_unit2.htm; South Carolina Electric and Gas, "Project Schedule," http://www.sceg.com/en/about-sceg/power-plants/new-nuclear development/schedule/default.htm; Southern Company, "Milestones," http://www. southerncompany.com/ nuclearenergy/milestones.aspx.

[7] Nuclear Regulatory Commission, "New Reactor Licensing Applications," September 10, 2012, http://www.nrc.gov/ reactors/new-reactors.html.

[8] World Nuclear Association, "Public Information Service," http://www.world-nuclear.org.

[9] Nuclear Regulatory Commission, "Actions in Response to the Japan Nuclear Accident: March 12, 2012," updated May 30, 2012, http://www.nrc.gov /reactors/operating/opsexperience/ japan/timeline/03122012.html.

[10] NRC, "Consideration of Additional Requirements for Containment Venting Systems for Boiling Water Reactors with Mark I and Mark II Containments," staff requirements memorandum, SECY-12-0157, March 19, 2013, http://www.nrc.gov/reading-rm/doc-collections /commission /srm/2012/2012-0157srm.pdf; Freebairn, William, "NRC Staff Recommends Ordering Filtered Vents for 31 Power Reactors," *Inside NRC*, November 5, 2012, p. 1.

[11] Gutherman Technical Services, "2013 Used Fuel Data," Report to Nuclear Energy Institute, January 20, 2014.

[12] DOE, *Strategy for the Management and Disposal of Used Nuclear Fuel and High-Level Radioactive Waste*, January 2013, http://energy. gov/sites/prod/files/2013%20115% 20Nuclear_Waste_Report.pdf.

[13] U.S. Court of Appeals for the District of Columbia Circuit, In re: Aiken County et al., No. 11-1271, writ of mandamus, August 13, 2013, http:// www.cadc.uscourts.gov/internet/ opinions.nsf/ BAE0CF34F762EBD985257BC6004DEB18/$file/11-1271-1451347.pdf.

[14] NRC, "Enhancements to Emergency Preparedness Regulations," final rule, *Federal Register*, November 23, 2011, p. 72560.

[15] NRC, "Request for Information Pursuant to Title 10 of the Code of Federal Regulations 50.54(f) Regarding Recommendations 2.1, 2.3, and 9.3 of the Near-Term Task Force Review of Insights from the Fukushima Dai-ichi Accident," March 12, 2012, http://pbadupws. nrc.gov/ docs/ML1205 /ML12053A340.pdf.

[16] World Nuclear Association, "World Nuclear Power Reactors & Uranium Requirements," February 1, 2014, http://www.world-nuclear.org /info/ reactors.html.

In: Challenges Facing the Nuclear Power Industry ISBN: 978-1-63483-618-0
Editor: Renee Olson

Chapter 3

CIVILIAN NUCLEAR WASTE DISPOSAL*

Mark Holt

SUMMARY

Management of civilian radioactive waste has posed difficult issues for Congress since the beginning of the nuclear power industry in the 1950s. Federal policy is based on the premise that nuclear waste can be disposed of safely, but proposed storage and disposal facilities have frequently been challenged on safety, health, and environmental grounds. Although civilian radioactive waste encompasses a wide range of materials, most of the current debate focuses on highly radioactive spent fuel from nuclear power plants. The United States currently has no disposal facility for spent nuclear fuel.

The Nuclear Waste Policy Act of 1982 (NWPA) calls for disposal of spent nuclear fuel in a deep geologic repository. NWPA established the Office of Civilian Radioactive Waste Management (OCRWM) in the Department of Energy (DOE) to develop such a repository, which would be licensed by the Nuclear Regulatory Commission (NRC). Amendments to NWPA in 1987 restricted DOE's repository site studies to Yucca Mountain in Nevada. DOE submitted a license application for the proposed Yucca Mountain repository to NRC on June 3, 2008. The state of Nevada strongly opposes the Yucca Mountain project, citing excessive

* This is an edited, reformatted and augmented version of a Congressional Research Service publication RL33461, prepared for Members and Committees of Congress dated April 24, 2015.

water infiltration, earthquakes, volcanoes, human intrusion, and other technical issues.

The Obama Administration "has determined that developing the Yucca Mountain repository is not a workable option and the Nation needs a different solution for nuclear waste disposal," according to the DOE FY2011 budget justification. As a result, no funding for Yucca Mountain, OCRWM, or NRC licensing was requested or provided for FY2011 or subsequent years. NRC halted further consideration of the license application in 2011 because of "budgetary limitations," but a federal appeals court on August 13, 2013, ordered NRC to continue the licensing process with previously appropriated funds. NRC completed its safety evaluation report on Yucca Mountain on January 29, 2015.

After halting the Yucca Mountain project, the Administration established the Blue Ribbon Commission on America's Nuclear Future to develop an alternative nuclear waste policy. The commission issued its final report on January 26, 2012, recommending that a new, "single-purpose organization" be given the authority and resources to promptly begin developing one or more nuclear waste repositories and consolidated storage facilities. The commission recommended a "consent based" process for siting nuclear waste storage and disposal facilities and that long-term research, development, and demonstration be conducted on technologies that could provide waste disposal benefits.

After OCRWM was dismantled, responsibility for implementing the Administration's nuclear waste policy was given to DOE's Office of Nuclear Energy (NE). In January 2013, NE issued a nuclear waste strategy based on the Blue Ribbon Commission recommendations. The strategy calls for a pilot interim storage facility for spent fuel from closed nuclear reactors to open by 2021 and a larger storage facility, possibly at the same site, to open by 2025. A site for a permanent underground waste repository would be selected by 2026, and the repository would open by 2048. DOE requested $30 million for FY2016 to develop an integrated waste management system as outlined by the new waste strategy—up from $22.5 million provided for FY2015. The House Appropriations Committee on April 22, 2015, approved $175 million for DOE and NRC to continue the Yucca Mountain licensing process and provided no funding for DOE's integrated waste strategy (H.R. 2028, H.Rept. 114-91).

Most Recent Developments

The Obama Administration's nuclear waste policy calls for termination of the Yucca Mountain repository project and the development of alternative approaches to waste management. Under the Nuclear Waste Policy Act of

1982 (NWPA), the Yucca Mountain site in Nevada has been the only location under consideration by the Department of Energy (DOE) for construction of a national high-level radioactive waste repository. DOE had submitted a license application for the Yucca Mountain repository to the Nuclear Regulatory Commission (NRC) on June 3, 2008, but the Obama Administration halted funding for the project and moved to withdraw the application on March 3, 2010.

Yucca Mountain Licensing

Responding to a court order that the Yucca Mountain repository licensing process continue as long as previously appropriated funding was available, NRC issued the final two volumes of the Yucca Mountain Safety Evaluation Report (SER) on January 29, 2015. The SER contains the NRC staff's determination of whether the repository would meet all applicable standards. Volume 3 of the SER, issued in October 2014, concluded that DOE's Yucca Mountain repository design would comply with safety and environmental standards after being permanently sealed.[1]

However, the staff said upon completing the SER that NRC should not authorize construction of the repository until all land and water rights requirements were met and a supplement to DOE's environmental impact statement (EIS) was completed.[2] NRC ordered its staff on March 3, 2015, to complete the supplemental EIS and make its database of Yucca Mountain licensing documents publicly available, using all the remaining previously appropriated licensing funds.[3] NRC Chairman Stephen Burns testified March 4, 2015, that $330 million in additional appropriations would be needed to complete the licensing process, including adjudicatory hearings on as many as 300 issues that have been raised by the State of Nevada and others.[4]

Obama Nuclear Waste Strategy

President Obama authorized DOE on March 24, 2015, to begin planning a separate underground repository for high-level radioactive waste generated by nuclear defense activities. This reverses a 1985 decision by the Reagan Administration to dispose of defense and civilian waste together. Energy Secretary Ernest Moniz described the proposed defense-only repository as potentially easier to site, license, and construct than a combined defense-civilian repository, because defense waste constitutes a relatively small portion of total high-level waste volumes and radioactivity and that some defense waste is in forms that might be optimized for certain types of disposal, such as deep boreholes. Moniz also said DOE would begin implementing a process to locate voluntary sites for storage and disposal of civilian nuclear waste.[5]

In a report issued in October 2014, DOE concluded that a defense-only nuclear waste repository "could be sited and developed outside the framework of the Nuclear Waste Policy Act." Under this reasoning, NWPA would not have to be amended to allow a defense-only repository to proceed. However, according to the DOE report, "Any such repository would be subject to licensing by the U.S. Nuclear Regulatory Commission and would have to comply with other NWPA requirements related to state and local participation in the siting process."[6]

The Obama Administration's nuclear waste policy is based on recommendations by the Blue Ribbon Commission on America's Nuclear Future (BRC), which was established after the Administration decided to terminate the Yucca Mountain repository project. The commission issued its final report on January 26, 2012, recommending that a new, "single-purpose organization" be given the authority and resources to promptly begin developing one or more nuclear waste repositories and consolidated storage facilities. The recommendations called for a "consent based" process in which the roles of various levels of government in siting and regulating nuclear waste facilities would be established through negotiations. The commission also recommended that long-term research, development, and demonstration be conducted on technologies that could provide waste disposal benefits.

In response to the BRC report, and to provide an outline for a new nuclear waste program, DOE issued a *Strategy for the Management and Disposal of Used Nuclear Fuel and High-Level Waste* in January 2013. The DOE strategy calls for a new nuclear waste management entity to develop consent-based storage and disposal sites, similar to the BRC recommendation. Under the DOE strategy, a pilot interim spent fuel storage facility would be opened by 2021 and a larger-scale storage facility, which could be an expansion of the pilot facility, by 2025. A geologic disposal facility would open by 2048—50 years after the initially planned opening date for the Yucca Mountain repository.[7]

Although Congress has approved the Obama Administration's halt in Yucca Mountain funding, it has not amended NWPA, which still names Yucca Mountain as the sole candidate site for a national repository for civilian nuclear waste. The House for several years has approved funding to resume the Yucca Mountain licensing process, although it has been blocked by the Senate. Republican leaders of the House Committee on Energy and Commerce issued a statement on March 24, 2015, criticizing DOE's plan for a defense-only nuclear waste repository as a way to deflect efforts to resume progress on Yucca Mountain.[8]

Consent-Based Siting Legislation

Bipartisan legislation (S. 854) that would implement several of the major BRC recommendations was introduced March 24, 2015, by Senator Lamar Alexander, Chairman of the Senate Appropriations Subcommittee on Energy and Water Development. S. 854 would establish an independent Nuclear Waste Administration to find sites for nuclear waste storage and disposal facilities with the consent of state and local officials and affected Indian tribes.

Development work could begin immediately on a pilot storage facility for waste from decommissioned reactors and emergency waste shipments from operating reactors. Work on interim storage facilities for other nuclear waste could be conducted within the first 10 years after enactment as long as funding had been obligated for a permanent repository. After 10 years, additional storage facilities could not be sited unless a candidate site had been selected for a permanent repository.

After enactment, all nuclear waste fees paid by nuclear reactor operators would be held in a new working capital fund, which would be available to the Nuclear Waste Administration without further appropriation by Congress. Balances in the existing Nuclear Waste Fund would continue to be subject to appropriation. The Nuclear Waste Administration would be authorized to develop defense-only storage and disposal facilities if the Secretary of Energy determined that defense and commercial nuclear waste should be managed separately.

Co-sponsoring S. 854 were Senator Diane Feinstein, Ranking Democrat on the Energy and Water subcommittee, Senator Lisa Murkowski, Chairman of the Energy and Natural Resources Committee, and Senator Maria Cantwell, Ranking Democrat on the Energy and Natural Resources Committee. The bill is similar to S. 1240 in the 113th Congress.

Voluntary Waste Storage Site

The waste management company Waste Control Specialists (WCS) announced February 7, 2015, that it would apply for an NRC license to develop an interim storage facility for spent nuclear fuel in Texas. The spent fuel storage facility would be built at a 14,000-acre WCS site near Andrews, TX, where the company currently operates two low-level radioactive waste storage facilities with local support. WCS said it would submit the final license application for the interim storage facility by April 2016 and complete construction by December 2020.[9] Under the WCS proposal, DOE would take title to spent fuel at nuclear plant sites, ship it to the Texas site, and pay WCS for storage for as long as 40 years with possible extensions, according to media reports. DOE's costs would be covered through appropriations from the Nuclear Waste Fund, as were most costs for the Yucca Mountain project. WCS contends that a privately developed spent fuel storage facility would not be bound by NWPA restrictions that prohibit DOE from building a storage facility without making progress on Yucca Mountain.[10]

Waste Program Appropriations

President Obama's FY2016 budget, submitted to Congress February 2, 2015, included no funding for Yucca Mountain but requested $30 million for FY2016 to develop an integrated waste management system as outlined by the Administration's waste strategy—up from $22.5 million provided for FY2015. Integrated waste management activities would include "preliminary generic process development and other non-R&D activities related to storage, transportation, and consent based siting," according to the DOE justification. Of the $30 million requested for integrated waste management, $24 million would come from the nuclear waste fund.[11]

In addition to integrated waste management, DOE requested $75.4 million for used (spent) fuel research and development, an increase of $26.4 million from the FY2015 level. Planned R&D activities include storage cask testing, modeling of spent fuel behavior during storage and transportation, studies of deep borehole disposal, and analysis and testing of crystalline, clay/shale, and salt rock types for geologic disposal of nuclear waste.

The House Appropriations Committee approved an FY2016 Energy and Water Development Appropriations bill April 22, 2015, that would appropriate $175 million from the Nuclear Waste Fund to continue the Yucca Mountain licensing process—$150 million for DOE and $25 million for NRC (H.R. 2028, H.Rept. 114-91). No funding would be provided for DOE's integrated waste management program. The House had approved $205 million for Yucca Mountain in the FY2015 Energy and Water bill but the funding was dropped in the enacted version.

Nuclear Waste Fee Collections

DOE stopped collecting nuclear waste fees from nuclear power generators on May 16, 2014, pursuant to a court ruling.[12] Citing uncertainty about the future of the nuclear waste program, the U.S. Court of Appeals for the District of Columbia Circuit had ordered DOE on November 19, 2013, to stop collecting fees on nuclear power that are supposed to pay for waste disposal. The fees, authorized by NWPA, had been paid by nuclear power generators at the rate of a tenth of a cent per kilowatt-hour and totaled about $750 million per year. NWPA requires the Secretary of Energy to adjust the fees as necessary to cover the waste program's anticipated costs, but the Court ruled that DOE's current waste plans are too vague to allow a reasonable estimate to be calculated. The Court noted that DOE's most recent cost estimate for the program had an uncertainty range of nearly $7 trillion, a range "so large as to be absolutely useless" for determining the waste fee.[13]

Waste Confidence and Continued Storage Rulemaking

NRC approved a final rule August 26, 2014, on continued storage of spent nuclear fuel.[14] The continued storage rule takes the place of NRC's earlier "waste confidence" rule, which was struck down by the U.S. Court of Appeals for the District of Columbia on June 8, 2012. The waste confidence rule had spelled out NRC's formal findings that waste generated by nuclear power plants would be disposed of safely—specifically, that spent nuclear fuel could be safely stored at nuclear plants for at least 60 years after they had shut down and that permanent disposal would be available "when necessary." The court ruled that NRC should have conducted an environmental review under the National Environmental Policy Act before issuing the most recent waste confidence findings in December 2010.[15] Under previous court rulings, NRC must determine that waste from proposed nuclear plants can be safely managed before licensing them to operate. As a result, after the court struck down the waste confidence rule, NRC halted licensing of new facilities that would generate radioactive waste. In approving the continued storage rule in August 2014, NRC ended its suspension of final licensing decisions for new reactors, spent fuel storage facilities, and license renewals.

Spent Fuel Pool Safety

The March 11, 2011, disaster at Japan's Fukushima Daiichi nuclear power plant increased concern about spent fuel stored in pools of water at nuclear plant sites. To reduce the potential hazard of spent fuel storage pools, Senator Edward J. Markey introduced legislation (S. 945) April 15, 2015, to require nuclear power plants to develop NRC-approved plans for removing spent fuel from storage pools. Within seven years after such plans were submitted, spent fuel would have to be transferred to dry storage facilities. A bill introduced by Senator Barbara Boxer the same day (S. 944) would require NRC to maintain full safety and security requirements at permanently closed reactors until all their spent fuel was moved to dry storage.

The loss of power at the Fukushima plant, caused by a huge earthquake and tsunami, disabled cooling systems at the plant's spent fuel pools. Water in the pools was initially feared to have boiled or leaked and dropped below the level of the stored spent fuel, but later analysis indicated that the spent fuel did not overheat. Nevertheless, the incident has prompted numerous recommendations for safety improvements at spent fuel pools. NRC approved an order March 9, 2012, requiring U.S. reactors to install improved water-level monitoring equipment at their spent fuel pools.[16]

Proposed Low-Level Waste Regulations

NRC proposed a significant modification of its low-level waste disposal regulations on March 26, 2015.[17] Among the proposed changes are a requirement that technical analyses for disposal sites include a 1,000-year compliance period for the general public and a dose limit for inadvertent intruders. New analyses would be required for 10,000-year and post-10,000-year periods, along with annual dose minimization targets for the first 10,000 years.

INTRODUCTION

Nuclear waste has sometimes been called the Achilles' heel of the nuclear power industry. Much of the controversy over nuclear power centers on the lack of a disposal system for the highly radioactive spent fuel that must be regularly removed from operating reactors.[18] Low-level radioactive waste generated by nuclear power plants, industry, hospitals, and other activities is also a long-standing issue.

Spent Nuclear Fuel Program

The Nuclear Waste Policy Act (P.L. 97-425), as amended in 1987, required DOE to focus on Yucca Mountain, Nevada, as the site of a deep underground repository for spent nuclear fuel and other highly radioactive waste. The state of Nevada has strongly opposed DOE's efforts on the grounds that the site is unsafe, pointing to potential volcanic activity, earthquakes, water infiltration, underground flooding, nuclear chain reactions, and fossil fuel and mineral deposits that might encourage future human intrusion.

Under the George W. Bush Administration, DOE determined that Yucca Mountain was suitable for a repository and that licensing of the site should proceed, as specified by NWPA. DOE submitted a license application for the repository to NRC on June 3, 2008, and projected that the repository could begin receiving waste in 2020, about 22 years later than the 1998 goal established by NWPA.

However, the Obama Administration made a policy decision that the Yucca Mountain repository should not be opened, largely because of Nevada's continuing opposition, although it requested FY2010 funding to continue the NRC licensing process. But the Administration's FY2011 budget request reversed the previous year's plan to continue licensing the repository and called for a complete halt in funding and elimination of the Office of Civilian Radioactive Waste Management (OCRWM), which had run the program. In line with the request, the FY2011 Continuing Appropriations Act (P.L. 112-10) provided no DOE funding for the program. DOE shut down the Yucca Mountain project at the end of FY2010 and transferred OCRWM's remaining functions to the Office of Nuclear Energy.

DOE filed a motion to withdraw the Yucca Mountain license application on March 3, 2010, "with prejudice," meaning the application could not be resubmitted to NRC in the future.[19] DOE's motion to withdraw the license application, filed with NRC's Atomic Safety and Licensing Board (ASLB), received strong support from the state of Nevada but drew opposition from states with defense-related and civilian radioactive waste that had been expected to go to Yucca Mountain. State utility regulators also filed a motion to intervene on March 15, 2010, contending that "dismissal of the Yucca Mountain application will significantly undermine the government's ability to fulfill its outstanding obligation to take possession and dispose of the nation's spent nuclear fuel and high level nuclear waste."[20]

The ASLB denied DOE's license withdrawal motion June 29, 2010, ruling that the NWPA prohibits DOE from withdrawing the license application until NRC determines whether the repository is acceptable.[21] The NRC commissioners sustained the ASLB decision on a tie vote September 9, 2011. However, NRC halted further consideration of the license application because of "budgetary limitations."[22] Lawsuits to overturn the Yucca Mountain license withdrawal on statutory grounds were filed with the U.S. Court of Appeals for the District of Columbia Circuit, which ruled on August 13, 2013, that NRC must continue work on the Yucca Mountain license application as long as funding is available. The Court determined that NRC had at least $11.1 million in previously appropriated funds for that purpose.[23]

NRC responded November 18, 2013, by directing the agency's staff to complete the Yucca Mountain safety evaluation report (SER), a key document that would provide the staff's conclusions about whether the proposed repository could be licensed.[24] NRC issued Volume 3 of the SER on October 16, 2014, concluding that DOE's Yucca Mountain repository design would comply with safety and environmental standards for one million years after being permanently sealed.[25] NRC issued the final two volumes of the Yucca Mountain SER on January 29, 2015.[26]

Upon completing the SER, the staff said that NRC should not authorize construction of the repository until all land and water rights requirements were met and a supplement to DOE's environmental impact statement (EIS) was completed. NRC ordered its staff on March 3, 2015, to complete the supplemental EIS and make its database of Yucca Mountain licensing documents publicly available, using all the remaining previously appropriated licensing funds.[27] NRC Chairman Stephen Burns testified March 4, 2015, that $330 million in additional appropriations would be needed to complete the licensing process, including adjudicatory hearings on as many as 300 issues that have been raised by the State of Nevada and others.[28]

After halting the Yucca Mountain project in 2010, the Obama Administration established the Blue Ribbon Commission on America's

Nuclear Future (BRC) to develop alternative waste disposal strategies. The BRC issued its final report on January 26, 2012, recommending that a new, "single-purpose organization" be given the authority and resources to promptly begin developing one or more nuclear waste repositories and consolidated storage facilities. The new organization would use a "consent based" process to select waste facility sites.[29] The BRC had commissioned a series of reports on various aspects of nuclear waste policy to assist in its deliberations.[30]

In response to the BRC report, and to provide an outline for a new nuclear waste program, DOE issued its *Strategy for the Management and Disposal of Used Nuclear Fuel and High-Level Waste* in January 2013. The DOE strategy calls for a new nuclear waste management entity to develop consent-based storage and disposal sites, similar to the BRC recommendation. Under the DOE strategy, a pilot interim spent fuel storage facility would be opened by 2021 and a larger-scale storage facility, which could be an expansion of the pilot facility, by 2025. A geologic disposal facility would open by 2048—50 years after the initially planned opening date for the Yucca Mountain repository.[31]

The nuclear power industry supports completion of NRC's licensing review of Yucca Mountain along with the pursuit of alternative storage and disposal facilities. "The target date for opening of Yucca Mountain or an alternative repository site should be no more than 20 years after a consolidated storage site is opened," according to an industry policy statement.[32]

The safety of geologic disposal of spent nuclear fuel and high-level waste (HLW), as planned in the United States, depends largely on the characteristics of the rock formations from which a repository would be excavated. Because many geologic formations are believed to have remained undisturbed for millions of years, it appeared technically feasible to isolate radioactive materials from the environment until they decayed to safe levels. "There is strong worldwide consensus that the best, safest long-term option for dealing with HLW is geologic isolation," according to the National Research Council.[33]

However, as the Yucca Mountain controversy indicates, scientific confidence about the concept of deep geologic disposal has turned out to be difficult to apply to specific sites. Every high-level waste site that has been proposed by DOE and its predecessor agencies has faced allegations or discovery of unacceptable flaws, such as water intrusion or earthquake vulnerability, that could release unacceptable levels of radioactivity into the environment. Much of the problem results from the inherent uncertainty involved in predicting waste site performance for the 1 million years that nuclear waste is to be isolated under current regulations. Widespread public controversy has also arisen over potential waste transportation routes to the sites under consideration.

President Obama's budgets for FY2016 and previous years have included long-term research on a wide variety of technologies that could reduce the volume and toxicity of nuclear waste. The Bush Administration had proposed to demonstrate large-scale facilities to reprocess and recycle spent nuclear fuel by separating long-lived elements, such as plutonium, that could be made into new fuel and "transmuted" into shorter-lived radioactive isotopes. Spent fuel reprocessing, however, has long been controversial because of cost concerns and the potential weapons use of separated plutonium. The Obama Administration refocused DOE's nuclear waste research toward fundamental science and away from the near-term design and development of reprocessing facilities.

President Bush had recommended the Yucca Mountain site to Congress on February 15, 2002, and Nevada Governor Guinn submitted a notice of disapproval, or "state veto," April 8, 2002, as allowed by NWPA. The state veto would have blocked further repository development at Yucca Mountain if a resolution approving the site had not been passed by Congress and signed into law within 90 days of continuous session. An approval resolution was signed by President Bush July 23, 2002 (P.L. 107-200).[34]

Other Programs

Other types of civilian radioactive waste have also generated public controversy, particularly low-level waste, which is produced by nuclear power plants, medical institutions, industrial operations, and research activities. Civilian low-level waste currently is disposed of in large trenches at sites in the states of South Carolina, Texas, and Washington. However, the Washington facility does not accept waste from outside its region, and the South Carolina site is available only to the three members of the Atlantic disposal compact (Connecticut, New Jersey, and South Carolina) as of June 30, 2008. The lowest-concentration class of low-level radioactive waste (class A) is accepted by a Utah commercial disposal facility from anywhere in the United States.

Threats by states to close their disposal facilities led to congressional authorization of regional compacts for low-level waste disposal in 1985. The first, and so far only, new disposal site under the regional compact system opened on November 10, 2011, near Andrews, TX.[35] The Texas Legislature approved legislation in May 2011 to allow up to 30% of the facility's capacity to be used by states outside the Texas Compact, which consists of Texas and Vermont.[36]

NUCLEAR WASTE LITIGATION

NWPA Section 302 authorized DOE to enter into contracts with U.S. generators of spent nuclear fuel and other highly radioactive waste; under the contracts, DOE was to dispose of the waste in return for a fee on nuclear power generation. The act prohibited nuclear reactors from being licensed to operate without a nuclear waste disposal contract with DOE, and all reactor operators subsequently signed them. As required by NWPA, the "standard contract" specified that DOE would begin disposing of nuclear waste no later than January 31, 1998.[37]

After DOE missed the contractual deadline, nuclear utilities began filing lawsuits to recover their additional storage costs—costs they would not have incurred had DOE begun accepting waste in 1998 as scheduled. DOE reached its first settlement with a nuclear utility, PECO Energy Company (now part of Exelon), on July 19, 2000. The agreement allowed PECO to keep up to $80 million in nuclear waste fee revenues during the subsequent 10 years. However, other utilities sued DOE to block the settlement, contending that nuclear waste fees may be used only for the DOE waste program and not as compensation for missing the disposal deadline. The U.S. Court of Appeals for the 11th Circuit agreed, ruling September 24, 2002, that any compensation would have to come from general revenues or other sources than the waste fund. Subsequent nuclear waste compensation to utilities has come from the U.S. Treasury's Judgment Fund, a permanent account that is used to cover damage claims against the U.S. government. Payments from the Judgment Fund do not require appropriations.

Through FY2014, nuclear waste payments from the Judgment Fund included $3.2 billion from settlements and $1.3 billion from final court judgments, for a total of about $4.5 billion, according to DOE. By the end of FY2014, 33 lawsuits had been settled, representing utilities that generate 82% of U.S. nuclear electricity. Thirty-one cases had received final court judgments, and 19 cases remained pending.[38] Under the settlements, utilities submit annual reimbursement claims to DOE for any delay-related nuclear waste storage costs they incurred during that year. Any disagreements over reimbursable claims between DOE and a utility would go to arbitration.

Utilities that have not settled with the Department of Justice have continued seeking damage compensation through the U.S. Court of Federal Claims. Unlike the settlements, which cover all past and future damages resulting from DOE's nuclear waste delays, awards by the Court of Claims can cover only damages that have already been incurred; therefore, utilities must continue filing claims as they accrue additional delay-related costs.

DOE estimates that its potential liabilities for waste program delays could total as much as $27.1 billion, including the $4.5 billion already paid in settlements and final judgments.[39]

Delays in the federal waste disposal program could also lead to future environmental enforcement action over DOE's own high-level waste and spent fuel, mostly resulting from defense and research activities. Some of the DOE-owned waste is currently being stored in non-compliance with state and federal environmental laws, making DOE potentially subject to fines and penalties if the waste is not removed according to previously negotiated compliance schedules.

Nuclear Waste Fee Collections

The National Association of Regulatory Utility Commissioners (NARUC), representing state utility regulators, and the Nuclear Energy Institute (NEI), representing the nuclear industry, filed petitions with the U.S. Court of Appeals on April 2 and April 5, 2010, respectively, to halt the federal government's collection of fees on nuclear power under the NWPA contracts. The suits argued that the fees, totaling about $750 million per year, should not be collected while the federal government's nuclear waste disposal program has been halted.[40] DOE responded that the federal government still intends to dispose of the nation's nuclear waste and that the fees must continue to be collected to cover future disposal costs.[41] Energy Secretary Steven Chu issued a formal determination on November 1, 2010, that there was "no reasonable basis at this time" to conclude that excess funds were being collected for future nuclear waste disposal activities.[42]

The U.S. Court of Appeals for the District of Columbia Circuit ruled June 1, 2012, that Secretary Chu's determination that the nuclear waste fee should continue unchanged was not "a valid evaluation" and ordered him to conduct a more thorough study of the fee within six months. The court noted that the

Secretary's finding relied primarily on costs that had been projected for the Yucca Mountain site, which the Administration had terminated as "unworkable." The court concluded that the Secretary must evaluate the likely costs of reasonable alternatives and the timing of those costs, all of which would affect the level of nuclear waste fees required.[43]

DOE responded with a new fee adequacy assessment in January 2013 that evaluated the total costs of a variety of waste management scenarios. The costs of some scenarios exceeded projected revenues from the existing waste fee by as much as $2 trillion, but other scenarios resulted in a surplus of up to $5 trillion. Because of the widely varying results, DOE concluded that there was no clear evidence that the fee should be immediately raised or lowered.[44]

After NEI and NARUC asked for a review of DOE's latest fee adequacy assessment, the Circuit Court ordered DOE on November 19, 2013, to stop collecting the nuclear waste fees altogether. The Court ruled that DOE's current waste plans were too vague to allow a reasonable estimate to be calculated. The Court noted that DOE's $7 trillion uncertainty range for the program's cost was "so large as to be absolutely useless" for determining the waste fee.[45] Pursuant to the court ruling, DOE stopped collecting nuclear waste fees from nuclear power generators on May 16, 2014.[46]

License Withdrawal

DOE's motion to withdraw the Yucca Mountain license application "with prejudice," meaning that it could not be resubmitted in the future, was filed with NRC's Atomic Safety and Licensing Board (ASLB) on March 3, 2010. DOE's motion argued that the licensing process should be terminated because "the Secretary of Energy has decided that a geologic repository at Yucca Mountain is not a workable option" for long-term nuclear waste disposal. Subsequent DOE statements have reiterated that the license withdrawal motion was not based on scientific or technical findings. Instead, the policy change was prompted by the perceived difficulty in overcoming continued opposition from the state of Nevada and a desire to find a waste solution with greater public acceptance, according to DOE.[47] DOE contended that the license application should be withdrawn "with prejudice" because of the need to "provide finality in ending the Yucca Mountain project."[48]

The state of Nevada strongly endorsed DOE's motion to withdraw the license application with prejudice[49] and moved to intervene in a court challenge to the license withdrawal.[50] Nevada has long contended that the geology of the site is unsuitable for long-term nuclear waste disposal.

However, DOE's withdrawal motion drew opposition from states and localities with defense-related and civilian nuclear waste that had been expected to go to Yucca Mountain. The state of South Carolina, which has large amounts of high-level radioactive waste at DOE's Savannah River Site, and the state of Washington, which hosts extensive nuclear waste storage facilities at DOE's Hanford Site, filed motions to intervene in the Yucca Mountain licensing proceeding to oppose the license application withdrawal.

NARUC also filed a motion to intervene in the Yucca Mountain licensing proceedings, contending that "dismissal of the Yucca Mountain application will significantly undermine the government's ability to fulfill its outstanding obligation to take possession and dispose of the nation's spent nuclear fuel and high level nuclear waste." NARUC's motion also contended that $17 billion collected from utility ratepayers for the nuclear waste program would be wasted if the Yucca Mountain license application were withdrawn.[51] Also seeking to intervene were Aiken County, SC, and the Prairie Island Indian Community in Minnesota.

The ASLB denied DOE's license withdrawal motion June 29, 2010, ruling that NWPA prohibits DOE from withdrawing the license application until NRC determines whether the repository is acceptable. According to the board, "Surely Congress did not contemplate that, by withdrawing the Application, DOE might unilaterally terminate the Yucca Mountain review process in favor of DOE's independent policy determination that 'alternatives will better serve the public interest.'"[52]

In appealing the ASLB decision to the NRC commissioners, DOE argued in a July 9, 2010, brief that the Secretary of Energy has broad authority under the Atomic Energy Act and Department of Energy Organization Act "to make policy decisions regarding disposal of nuclear waste and spent nuclear fuel." DOE contended that such authority includes "the authority to discontinue the Yucca Mountain project" and that NRC rules provide "that applicants in NRC licensing proceedings may withdraw their applications."[53] After more than a year of deliberation, the NRC commissioners sustained the licensing board's decision on a tie vote September 9, 2011. However, NRC halted further consideration of the license application because of "budgetary limitations."[54]

South Carolina and Aiken County filed challenges to the Yucca Mountain license withdrawal in the U.S. Court of Appeals for the District of Columbia Circuit, contending that NWPA requires the licensing process to proceed.

After NRC rejected the license withdrawal motion, the plaintiffs in that case, including Nye County, NV, where Yucca Mountain is located, petitioned the court to order NRC to continue the licensing proceedings, as noted above.[55] The court held oral arguments in the case May 2, 2012. NRC filed an update with the court on January 4, 2013, contending that the reelection of President Obama and the exclusion of additional Yucca Mountain funding from the most recent FY2013 Continuing Resolution reinforced NRC's suspension of further action on the repository license.[56]

The Court of Appeals ruled on August 13, 2013, that NRC must continue work on the Yucca Mountain license application as long as funding is available. The Court determined that NRC had at least $11.1 million in previously appropriated funds for that purpose.[57] As noted above, NRC completed its Safety Evaluation Report for Yucca Mountain in January 2015 and will use the remaining funds to complete a supplemental EIS and make the licensing database available to the public. Beyond those actions, additional funding will be required, according to NRC.[58]

Waste Confidence Decision and Continued Storage Rule

Before issuing licenses to nuclear reactors and waste storage facilities, NRC is required by a 1979 court decision to determine that waste from those facilities can be safely disposed of.[59] To meet that requirement, NRC issued a Waste Confidence Decision in 1984 that found that nuclear waste could be safely stored at reactor sites for at least 30 years after plant closure and that a permanent repository would be available by 2007-2009.[60] At that time, DOE officially planned to meet the NWPA repository deadline of 1998.

After DOE's schedule for opening a nuclear waste repository began to slip, NRC updated the Waste Confidence Decision in 1990 to find that a repository would be available by the first quarter of the next century.[61] When the Yucca Mountain repository was delayed further and then terminated by the Obama Administration, NRC issued another waste confidence rule in 2010 that found that a repository would be available "when necessary" and that waste could be safely stored at reactor sites for at least 60 years after shutdown.[62]

The state of New York, environmental groups, and others filed lawsuits to overturn the 2010 waste confidence rule on the grounds that NRC had not adequately considered the environmental risks of long-term waste storage at reactor sites. The U.S. Court of Appeals for the District of Columbia Circuit largely agreed, ruling on June 8, 2012, that NRC would have to conduct an environmental review of the Waste Confidence Decision under the National Environmental Policy Act (NEPA). The court found two major flaws in NRC's rulemaking process:

> First, in concluding that permanent storage will be available "when necessary," the Commission did not calculate the environmental effects of failing to secure permanent storage—a possibility that cannot be ignored. Second, in determining that spent fuel can safely be stored on site at nuclear plants for sixty years after the expiration of a plant's license, the Commission failed to properly examine future dangers and key consequences.[63]

Final licensing of new facilities that would produce nuclear waste was halted for more than two years while NRC worked on its response to the court ruling. NRC approved a final rule August 26, 2014, on continued storage of spent nuclear fuel to replace the waste confidence rule that had been struck down.[64] Rather than make specific findings about the future availability of waste disposal facilities, the new continued storage rule describes environmental effects that may result from various periods of waste storage, based on the findings of a generic environmental impact statement (GEIS). The GEIS, issued along with the continued storage rule, responded to the court requirement for NEPA review.

The GEIS analyzed the environmental effects of three potential time periods of storage before a permanent repository would become available: "short-term timeframe," continued storage for up to 60 years after a reactor ceases operation; "long-term timeframe," for up to 160 years after reactor shutdown; and an "indefinite timeframe," in which a repository may never become available. The GEIS assumed that active management and oversight of the stored spent fuel would never end, and that "spent fuel canisters and casks would be replaced approximately once every 100 years." The environmental impact of all three time frames was judged to be minimal in almost all categories.[65]

NRC's continued storage rule was criticized by a coalition of environmental and other groups for allowing licensing of new nuclear reactors to continue while permanent disposal of their resulting waste remained uncertain.[66] The states of Connecticut, New York, and Vermont filed a federal lawsuit October 24, 2014, to vacate the continued storage rule and require NRC to prepare a new environmental impact statement.[67]

CONGRESSIONAL ACTION

President Obama's proposal to terminate the Yucca Mountain project and search for disposal alternatives, as outlined by DOE's January 2013 nuclear waste strategy, has prompted substantial congressional debate and a number of legislative proposals. The House has consistently opposed the Obama Administration's efforts to abandon Yucca Mountain, while the Senate has generally expressed more interest in alternative waste management proposals. However, the dynamics of the nuclear waste debate could be affected by the change in Senate control in the 114th Congress.

Senator Lamar Alexander, along with Senators Lisa Murkowski, Diane Feinstein, and Maria Cantwell, introduced legislation March 24, 2015, to redirect the nuclear waste program along the lines recommended by the Blue Ribbon Commission and the DOE waste strategy (S. 854). The bill, similar to S. 1240 in the 113th Congress, would establish an independent Nuclear Waste Administration (NWA) to develop nuclear waste storage and disposal facilities. Siting of such facilities would require the consent of the affected state, local, and tribal governments.

Under S. 854, NWA would be required to prepare a mission plan to open a pilot storage facility by the end of 2021 for nuclear waste from shutdown reactors and other emergency deliveries (called "priority waste"). A storage facility for waste from operating reactors or other "nonpriority waste" would open by the end of 2025, and a permanent repository by the end of 2048.

NWA would be authorized to issue requests for proposals or select sites for storage facilities for nonpriority waste only if, during the first 10 years after enactment, the agency had obligated funds for developing a permanent waste repository. After 10 years, NWA could not request proposals for nonpriority waste or select sites unless a candidate site for a repository had been selected. NWA would be authorized to offer financial compensation and other incentives for hosting nuclear waste storage and disposal facilities. Sites that would include storage facilities along with a repository would receive preference.

Highly radioactive defense waste, which had been planned for commingling with commercial nuclear waste since the 1980s, could be placed in defense-only storage and disposal facilities under S. 854, subject to appropriations. President Obama authorized DOE to pursue a defense-only repository on March 24, 2015.

Nuclear waste fees collected after enactment of the bill would be held in a newly established Working Capital Fund. The Nuclear Waste Administration could immediately draw from that fund any amounts needed to carry out S. 854, unless limited by annual appropriations or authorizations. The current disposal limit of 70,000 metric tons for the first repository under NWPA would be repealed.

The Senate Committee on Energy and Natural Resources held a hearing on S. 1240 on July 30, 2013. Energy Secretary Ernest J. Moniz, who had been a member of the Blue Ribbon Commission, said the bill provided "a promising framework for addressing key issues." NARUC Electricity Committee Chairman David C. Boyd called the bill "a step in the right direction," but urged that it require continued licensing action on the Yucca Mountain repository. Boyd noted that S. 1240 would not preclude enforcement of existing NWPA deadlines for action on Yucca Mountain. Natural Resources Defense Council Senior Attorney Geoffrey H. Fettus opposed the bill on the grounds that it would allow temporary waste storage facilities to be opened without progress on a permanent repository and that states would have inadequate authority to regulate repository safety, among other concerns.[68] As noted above, S. 854 would require certain actions on a permanent repository before a storage facility for nonpriority waste could be sited.

A pilot program to develop one or more voluntary interim storage sites for nuclear waste was included in the FY2014 Energy and Water Development appropriations bill passed by the Senate Appropriations Committee June 27, 2013 (S. 1245, §309). Before moving forward with any proposed waste storage facility, DOE would have to reach siting agreements with the governor of the proposed host state, each unit of local government with jurisdiction over the facility, and each affected Indian tribe. Such agreements would have to be approved by Congress, which would then appropriate funding for the project. Similar language was included by the committee in its FY2013 measure (S. 2465, §312) and in the draft Senate FY2015 Energy and Water bill approved in subcommittee.[69] Corresponding House appropriations bills have not included such an authorization, and it has not been enacted.

The debate over nuclear waste policy was strongly affected by the March 11, 2011, Fukushima Daiichi nuclear accident in Japan. The loss of power at the Fukushima site, caused by a huge earthquake and tsunami, disabled cooling systems at the plant's spent fuel pools. Water in the pools was initially suspected to have boiled or leaked and dropped below the level of the stored spent fuel, potentially leading to fuel damage and radioactive releases into the atmosphere. However, later analysis indicated that the spent fuel did not overheat.

Concerns have been raised in Congress about the risk posed by stored spent fuel, particularly that the cancellation of the Yucca Mountain repository would leave growing amounts of spent fuel indefinitely stored at nuclear plant sites throughout the United States. To reduce the potential hazard of spent fuel storage pools, Senator Edward J. Markey introduced legislation (S. 945) April 15, 2015, to require nuclear power plants to develop NRC-approved plans for removing spent fuel from storage pools. Within seven years after such plans had been submitted, spent fuel would have to be transferred to dry storage facilities. After the seven-year period, additional spent fuel would have to be transferred to dry casks within a year after it had been determined to be sufficiently cool. Emergency planning zones would have to be expanded from 10 to 50 miles in radius around any reactor determined by NRC to be out of compliance with its spent fuel transfer plan. The emergency zone for a decommissioned reactor could not be reduced below a 10-mile radius until all its spent fuel had been placed in dry storage. NRC would be authorized to use interest earned by the Nuclear Waste Fund to provide grants to nuclear power plants to transfer spent fuel to dry storage. Markey introduced a similar bill (S. 2325) in the 113th Congress.

A bill introduced by Senator Barbara Boxer also on April 15, 2015 (S. 944), similar to S. 2324 in the 113th Congress, would require NRC to maintain full safety and security requirements at permanently closed reactors until all their spent fuel was moved to dry storage.

NRC released a study on November 12, 2013, concluding that "expedited transfer of spent fuel to dry cask storage would provide only a minor or limited safety benefit" and "its expected implementation costs would not be warranted."[70]

CHARACTERISTICS AND HANDLING OF NUCLEAR WASTE

Radioactive waste is a term that encompasses a broad range of material with widely varying characteristics. Some waste has relatively slight radioactivity and is safe to handle in unshielded containers, while other types are intensely hot in both temperature and radioactivity. Some decays to safe levels of radioactivity in a matter of days or weeks, while other types will remain dangerous for thousands of years. Major types of radioactive waste are described below:[71]

Spent nuclear fuel. Fuel rods that have been withdrawn from a nuclear reactor after irradiation, usually because they can no longer efficiently sustain a nuclear chain reaction. (The term "spent nuclear fuel" is defined in NWPA. The nuclear industry typically refers to spent fuel as "used nuclear fuel," because it contains uranium and plutonium that could be extracted through reprocessing to make new fuel.) By far the most radioactive type of civilian nuclear waste, spent fuel contains extremely hot but relatively short-lived fission products (fragments of the nuclei of uranium and other fissile elements) as well as long-lived radionuclides (radioactive atoms) such as plutonium, which remains dangerously radioactive for tens of thousands of years or more.

High-level waste. Highly radioactive residue created by spent fuel reprocessing (almost entirely for defense purposes in the United States). High-level waste contains most of the radioactive fission products of spent fuel, but most of the uranium and plutonium usually has been removed for re-use. Enough long-lived radioactive elements typically remain, however, to require isolation for 10,000 years or more.

Transuranic (TRU) waste. Relatively low-activity waste that contains more than a certain level of long-lived elements heavier than uranium (primarily plutonium). Radiation shielding may be required for the handling of some types of TRU waste. In the United States, transuranic waste is generated almost entirely by nuclear weapons production processes. Because of the plutonium, long-term isolation is required. The nation's only permanent repository for TRU waste, the Waste Isolation Pilot Plant (WIPP), near Carlsbad, NM, suspended disposal operations after a radioactive release on February 14, 2014, and plans to resume in early 2016.[72]

Low-level waste. Radioactive waste not classified as spent fuel, high-level waste, TRU waste, or byproduct material such as uranium mill tailings (below). Four classes of low-level waste have been established by NRC, ranging from least radioactive and shortest-lived to the longest-lived and most radioactive. Although some types of low-level waste can be more radioactive than some types of high-level waste, in general low-level waste contains relatively low concentrations of radioactivity that decays relatively quickly. Low-level waste disposal facilities cannot accept material that exceeds NRC concentration limits.

Uranium mill tailings. Sand-like residues remaining from the processing of uranium ore. Such tailings have very low radioactivity but extremely large volumes that can pose a hazard, particularly from radon emissions or groundwater contamination.

Mixed waste. Chemically hazardous waste that includes radioactive material. High-level, low-level, and TRU waste, and radioactive byproduct material, often falls under the designation of mixed waste. Such waste poses complicated institutional problems, because the radioactive portion is regulated by DOE or NRC under the Atomic Energy Act, while the Environmental Protection Agency (EPA) and states regulate the non-radioactive elements under the Resource Conservation and Recovery Act (RCRA).

Spent Nuclear Fuel

When spent nuclear fuel is removed from a reactor, usually after several years of power production, it is thermally hot and highly radioactive. The spent fuel is in the form of fuel assemblies, which consist of arrays of metal-clad fuel rods 12-15 feet long.

A fresh fuel rod, which emits relatively little radioactivity, contains uranium that has been enriched in the isotope U-235 (usually to 3%-5% from its natural level of 0.7%). But after nuclear fission has taken place in the reactor, most of the U-235 nuclei in the fuel rods have been split into a variety of highly radioactive fission products. Some of the nuclei of the dominant isotope U-238 have absorbed neutrons and then decayed to become radioactive plutonium, some of which has also split into fission products (and some of which are gases). Newly withdrawn spent fuel assemblies are stored in deep pools of water adjacent to the reactors to keep them from overheating and to protect workers from radiation. To prevent the pools from filling up, older, cooler spent fuel often is sealed in dry canisters and transferred to radiation-shielded storage facilities elsewhere at reactor sites. NRC currently requires spent fuel to cool for at least 7-10 years before being transferred to dry storage.[73]

Spent fuel discharged from U.S. commercial nuclear reactors is currently stored at 60 operating nuclear plant sites, 14 shutdown plant sites, and the Idaho National Laboratory.[74] A typical large commercial nuclear reactor discharges an average of 20-30 metric tons of spent fuel per year—an average of about 2,150 metric tons annually for the entire U.S. nuclear power industry. The nuclear industry estimated that the total amount of commercial spent fuel was 71,775 metric tons at the end of 2013, including 22,233 metric tons in dry storage and other separate storage facilities.[75] The total amount of existing waste would exceed NWPA's 70,000-metric-ton limit for Yucca Mountain, even without counting 7,000 metric tons of DOE spent fuel and high-level waste that had also been planned for disposal at the repository.

As long as nuclear power continues to be generated, the amount of spent fuel stored at plant sites will continue to grow until an interim storage facility or a permanent repository can be opened— or until alternative treatment and disposal technology is developed. DOE's most recent estimates of the total amount of U.S. commercial spent fuel that may eventually require disposal range from 105,000 metric tons[76] to 130,000 metric tons.[77]

New storage capacity at operating nuclear plant sites or other locations will be required if DOE is unable to begin accepting waste into its disposal system for an indefinite period. Most utilities are expected to construct new dry storage capacity at reactor sites. Sixty-two licensed dry storage facilities were operating at nuclear plant sites in the United States at the end of 2013.[78]

The terrorist attacks of September 11, 2001, heightened concerns about the vulnerability of stored spent fuel. Concerns have been raised that an aircraft crash into a reactor's pool area or acts of sabotage could drain the pool and cause the spent fuel inside to overheat. A report released by NRC January 17, 2001, found that overheating could cause the zirconium alloy cladding of spent fuel to catch fire and release hazardous amounts of radioactivity, although it characterized the probability of such a fire as low.

In a report released April 6, 2005, the National Academy of Sciences (NAS) found that "successful terrorist attacks on spent fuel pools, though difficult, are possible." To reduce the likelihood of spent fuel cladding fires, the NAS study recommended that hotter and cooler spent fuel assemblies be interspersed throughout spent fuel pools, that spray systems be installed above the pools, and that more fuel be transferred from pools to dry cask storage.[79] The nuclear industry contends that the several hours required for uncovered spent fuel to heat up enough to catch fire would allow ample time for alternative measures to cool the fuel. NRC's report on this issue in 2013 found only minor safety benefits in expedited transfers of spent fuel from pools to dry casks.[80]

The safety of spent fuel pools is one of the areas examined by an NRC task force that identified near-term lessons that the Fukushima accident may hold for U.S. nuclear power plant regulation. The task force recommended that assured sources of electrical power as well as water spray systems be available for spent fuel pools.[81] NRC approved an order March 9, 2012, requiring U.S. reactors to install improved water-level monitoring equipment at their spent fuel pools.[82]

Commercial Low-Level Waste

About 1.2 million cubic feet of low-level waste with about 17,217 curies of radioactivity was shipped to commercial disposal sites in 2013, according to DOE.[83] Volumes and radioactivity can vary widely from year to year, based on the status of nuclear decommissioning projects and cleanup activities that can generate especially large quantities. The radioactivity of low-level waste is only a tiny fraction of the amount in annual discharges of spent fuel.

Low-level radioactive waste is divided into three major categories for handling and disposal: Class A, B, and C. Class A waste constitutes most of the annual volume of low-level waste, while classes B and C generally contain most of the radioactivity. As discussed below, most of the nation's Class B and C waste has been stored where it has been generated since June 2008 for lack of a permanent disposal site.

CURRENT POLICY AND REGULATION

Disposal of spent fuel and high-level waste is a federal responsibility, while states are authorized to develop disposal facilities for commercial low-level waste. The Obama Administration has halted the Yucca Mountain repository, although it remains the sole candidate site for civilian highly radioactive waste disposal under current law. DOE issued an alternative waste management strategy in January 2013 that calls for a pilot facility for spent fuel storage to open at a voluntary site by 2021 and a new repository by 2048. New legislation would be required to carry out the strategy.

Spent Nuclear Fuel

Current Program and Proposed Policy Changes

The Nuclear Waste Policy Act established a system for selecting a geologic repository for the permanent disposal of up to 70,000 metric tons (77,000 tons) of spent nuclear fuel and high-level waste. DOE's Office of Civilian Radioactive Waste Management (OCRWM) was created to carry out the program. The Nuclear Waste Fund, holding receipts from a fee on commercial nuclear power) and federal contributions for emplacement of high-level defense waste, was established to pay for the program. The fee, set at a tenth of a cent per kilowatt-hour, can be adjusted by the Secretary of

Energy based on projected total program costs after a congressional review period. DOE was required to select three candidate sites for the first national high-level waste repository.

After much controversy over DOE's implementation of NWPA, the act was substantially modified by the Nuclear Waste Policy Amendments Act of 1987 (Title IV, Subtitle A of P.L. 100- 203, the Omnibus Budget Reconciliation Act of 1987). Under the amendments, the only candidate site DOE may consider for a permanent high-level waste repository is at Yucca Mountain, Nevada. If that site cannot be licensed, DOE must return to Congress for further instructions.

The 1987 amendments also authorized construction of a monitored retrievable storage (MRS) facility to store spent fuel and prepare it for delivery to the repository. But because of fears that the MRS would reduce the need to open the permanent repository and become a de facto repository itself, the law forbids DOE from selecting an MRS site until recommending to the President that a permanent repository be constructed, and construction of an MRS cannot begin until Yucca Mountain receives a construction permit. The repository recommendation was made in February 2002, but DOE has not announced any plans for siting an MRS.

Along with halting all funding for the Yucca Mountain project, the Obama Administration terminated OCRWM at the end of FY2010 and transferred its remaining functions to DOE's Office of Nuclear Energy. The Administration established the Blue Ribbon Commission on America's Nuclear Future (BRC) to develop a new waste management strategy, and the BRC issued its final report on January 26, 2012.[84]

As required by its charter, the BRC did not evaluate specific sites for new nuclear waste facilities, including Yucca Mountain. However, the commission concluded that the existing nuclear waste policy, with Yucca Mountain identified by law as the sole candidate site, "has now all but completely broken down" and "seems destined to bring further controversy, litigation, and protracted delay." The BRC recommended instead that Congress establish "a new, consent-based approach to siting." Under that approach, potential sites would be the subject of extensive negotiations with affected states, tribes, and local governments. Such negotiations would result in legally binding agreements on the roles of the affected parties, including local oversight, and other project parameters.

The BRC noted that previous U.S. efforts to find voluntary waste sites had failed, but it nevertheless expressed confidence that such a process could eventually succeed. In particular, the commission highlighted the U.S.

experience with the Waste Isolation Pilot Plant (WIPP) in New Mexico, which, after many years of controversy, began receiving transuranic defense waste in 1999 with state and local government approval.

To carry out the new waste management program, the BRC recommended that a congressionally chartered federal corporation be established. Such a corporation would be independent from Administration control and have "assured access to funds" but be subject to congressional oversight and to regulation by NRC. Pending establishment of the corporation, the BRC recommended that administrative and legislative changes be implemented in the Nuclear Waste Fund to allow funds to be used for the waste management program without having to compete with other appropriations priorities.

The BRC called for "prompt efforts" to develop a permanent underground nuclear waste repository and to develop one or more interim central storage facilities. Interim storage facilities are especially needed so that waste can be removed from shutdown reactor sites, the commission said. Development of a permanent disposal site would have to be undertaken along with the interim storage effort to assure that interim sites would not become "de facto" permanent repositories, according to the commission.

In response to the BRC report, and to provide an outline for a new nuclear waste program, DOE issued its *Strategy for the Management and Disposal of Used Nuclear Fuel and High-Level Waste* in January 2013.[85] Under the DOE strategy, a pilot interim spent fuel storage facility would be opened by 2021, focusing primarily on spent fuel from decommissioned nuclear plants. A larger-scale interim storage facility, which could be an expansion of the pilot facility, would open by 2025 with a capacity of 20,000 metric tons or more.

The DOE strategy called for the interim storage facility to be linked to development of a permanent repository so that the storage facility would not become a *de facto* repository. However, the strategy noted that the existing NWPA restrictions on the MRS are so rigid that the MRS cannot currently be built. Without describing specific provisions, the DOE strategy recommended that "this linkage should not be such that it overly restricts forward movement on a pilot or larger storage facility that could make progress against the waste management mission."

Under the DOE strategy, a geologic disposal facility would open by 2048—50 years after the initially planned opening date for the Yucca Mountain repository. A site for the repository is to be selected by 2016, and site suitability studies, design, and licensing are to be completed by 2042.

Sites for the proposed storage and disposal facilities would be selected through a "consent based" process, as recommended by the BRC. However, the DOE strategy included few details on how such a process would be implemented. Instead, the strategy said the Obama Administration would soon begin consultations with Congress and interest groups on "defining consent, deciding how that consent is codified, and determining whether or how it is ratified by Congress." The program would be implemented by a new nuclear waste management entity, as recommended by the BRC, but the nature of the new organization was not specified by the DOE strategy. A bill introduced by Senator Alexander (S. 854), discussed under "Congressional Action," would modify the waste program along the lines of the Administration's waste strategy. Other proposals have called for privatization of waste management services.[86]

DOE issued a report in October 2014 that recommended testing the consent-based approach by siting and developing a repository solely for defense and research waste. According to the report, a separate repository for such waste would not be subject to the Yucca Mountain siting requirement that applies to a civilian nuclear waste repository under NWPA. The idea would reverse long-standing federal policy, established by the Reagan Administration, that a single repository would hold both civilian and defense high-level waste and spent fuel. DOE's 2014 report concluded that a separate repository for the nation's relatively small volumes of defense and research waste (compared to civilian waste) could be developed more quickly, "within existing legislative authority," than a repository for all highly radioactive waste. The report also recommended that disposal in deep boreholes be considered for the most compact types of defense and research waste.[87]

President Obama authorized DOE on March 24, 2015, to begin planning a separate underground repository for high-level radioactive waste generated by nuclear defense activities. Energy Secretary Moniz said at the same time that DOE would begin implementing a process to locate voluntary sites for storage and disposal of civilian nuclear waste.[88]

Private Interim Storage

The waste management company Waste Control Specialists (WCS) announced February 7, 2015, that it would apply for an NRC license to develop an interim storage facility for spent nuclear fuel in Texas. The spent fuel storage facility would be built at a 14,000-acre WCS site near Andrews, TX, where the company currently operates two low-level radioactive waste storage facilities with local support. WCS said it would submit the final license application for the interim storage facility by April 2016 and complete construction by December 2020.[89] Under the WCS proposal, DOE would take title to spent fuel at nuclear plant sites, ship it to the Texas site, and pay WCS for storage for up to 40 years with possible extensions, according to media reports. DOE's costs would be covered through appropriations from the Nuclear Waste Fund, as were most costs for the Yucca Mountain project. WCS contends that a privately developed spent fuel storage facility would not be bound by NWPA restrictions that prohibit DOE from building a storage facility without making progress on Yucca Mountain.[90]

Local officials near the WIPP facility have long supported the development of additional waste facilities at that site, which was originally planned to hold high-level waste before the state objected. A presentation by a top New Mexico official on March 1, 2012, described conditions under which the state might be willing to accept high-level waste and spent fuel at the WIPP site, such as assistance with cleaning up the state's contaminated uranium production sites.[91] A local government consortium near the WIPP site, the Eddy-Lea Energy Alliance, sent a letter to NRC on February 26, 2013, saying that it would submit a license application for the consolidated spent fuel storage facility envisioned by DOE's waste strategy report. "As details of the DOE strategy are implemented, we will keep the NRC staff advised of our progress," the letter said.[92] However, a February 2014 radioactive release from WIPP, which led to the suspension of disposal operations, could affect New Mexico's support for expanded waste activities. Interest in hosting nuclear waste sites has also been expressed by groups in Mississippi and Loving County, Texas, although whether they would be developed by the private sector or the government has not been specified.[93]

An earlier effort to develop a private spent fuel storage facility was undertaken after it became apparent that DOE would miss the 1998 deadline for taking nuclear waste from reactor sites. A utility consortium signed an agreement with the Skull Valley Band of the Goshute Indians in Utah on December 27, 1996, to develop a storage facility on tribal land. The Private Fuel Storage (PFS) consortium submitted a license application to NRC on June 25, 1997, and a 20-year license for storing up to 44,000 tons of spent fuel in dry casks was issued on February 21, 2006. However, NRC noted that Interior Department approval would also be required.

On September 7, 2006, the Department of the Interior issued two decisions against the PFS project. The Bureau of Indian Affairs disapproved a proposed lease of tribal trust lands to PFS, concluding there was too much risk that the waste could remain at the site indefinitely.[94] The Bureau of Land Management rejected the necessary rights-of-way to transport waste to the facility, concluding that a proposed rail line would be incompatible with the

Cedar Mountain Wilderness Area and that existing roads would be inadequate.[95]

The Skull Valley Band of Goshutes and PFS filed a federal lawsuit July 17, 2007, to overturn the Interior decisions on the grounds that they were politically motivated.[96] A federal district court judge on July 26, 2010, ordered the Department of the Interior to reconsider its decisions on the PFS permits.[97] However, PFS asked NRC to terminate its license on December 20, 2012.[98]

Regulatory Requirements for Yucca Mountain

Although the Obama Administration wants to redirect the high-level nuclear waste program, current law still focuses on Yucca Mountain for civilian waste. NWPA requires that high-level waste repositories be licensed by NRC in accordance with general standards issued by EPA. Under the Energy Policy Act of 1992 (P.L. 102-486), EPA was required to write new repository standards specifically for Yucca Mountain. NWPA also requires the repository to meet general siting guidelines prepared by DOE and approved by NRC. Transportation of waste to storage and disposal sites is regulated by NRC and the Department of Transportation (DOT). Under NWPA, DOE shipments to Yucca Mountain and an MRS facility would have to use NRC-certified casks and comply with NRC requirements for notifying state and local governments. Shipments would also have to follow DOT regulations on routing, placarding, and safety.

NRC's licensing requirements for Yucca Mountain, at 10 C.F.R. 63, require compliance with EPA's standards (described below) and establish procedures that DOE must follow in seeking a repository license. For example, DOE is required to conduct a repository performance confirmation program that would indicate whether natural and man-made systems were functioning as intended and assure that other assumptions about repository conditions were accurate.

Specific standards for Yucca Mountain were required because of concerns that some of EPA's general standards might be impossible or impractical to meet at Yucca Mountain.[99] The Yucca Mountain standards, which limit the radiation dose that the repository could impose on individual members of the public, were required to be consistent with the findings of a study by the National Academy of Sciences (NAS), which was issued August 1, 1995.[100] The NAS study recommended that the Yucca Mountain environmental standards establish a limit on risk to individuals near the repository, rather than setting specific limits for the releases of radioactive material or on radioactive doses, as under previous EPA standards. The NAS study also examined the potential for human intrusion into the repository and found no scientific basis for predicting human behavior thousands of years into the future.

Pursuant to the Energy Policy Act of 1992, EPA published its proposed Yucca Mountain radiation protection standards on August 27, 1999. The proposal would have limited annual radiation doses to 15 millirems for the "reasonably maximally exposed individual," and to 4 millirems from groundwater exposure, for the first 10,000 years of repository operation. EPA calculated that its standard would result in an annual risk of fatal cancer for the maximally exposed individual of seven chances in a million. The nuclear industry criticized the EPA proposal as being unnecessarily stringent, particularly the groundwater standard. On the other hand, environmental groups contended that the 10,000-year standard proposed by EPA was too short, because DOE had projected that radioactive releases from the repository would peak after about 400,000 years.

EPA issued its final Yucca Mountain standards on June 6, 2001. The final standards included most of the major provisions of the proposed version, including the 15 millirem overall exposure limit and the 4 millirem groundwater limit. Despite the department's opposition to the EPA standards, DOE's site suitability evaluation determined that the Yucca Mountain site would be able to meet them. NRC revised its repository regulations September 7, 2001, to conform to the EPA standards.

A three-judge U.S. Court of Appeals panel on July 9, 2004, struck down the 10,000-year regulatory compliance period in the EPA and NRC Yucca Mountain standards.[101] The court ruled that the 10,000-year period was inconsistent with the NAS study on which the Energy Policy Act required the Yucca Mountain regulations to be based. In fact, the court found, the NAS study had specifically rejected a 10,000-year compliance period because of analysis that showed peak radioactive exposures from the repository would take place several hundred thousand years in the future.

In response to the court decision, EPA proposed a new version of the Yucca Mountain standards on August 9, 2005. The proposal would have retained the dose limits of the previous standard for the first 10,000 years but allowed a higher annual dose of 350 millirems for the period of 10,000 years through 1 million years. EPA also proposed to base the post-10,000-year Yucca Mountain standard on the median dose, rather than the mean, potentially making it easier to meet.[102] Nevada state officials called EPA's proposed standard far too lenient and charged that it was "unlawful and arbitrary."[103]

EPA issued its final rule to amend the Yucca Mountain standards on September 30, 2008. The final rule reduced the annual dose limit during the period of 10,000 through 1 million years from the proposed 350 millirems to 100 millirems, which the agency contended was consistent with international standards. Under the final rule, compliance with the post-10,000-year standard will be based on the arithmetic mean of projected doses, rather than the median as proposed. The 4 millirem groundwater standard will continue to apply only to the first 10,000 years.[104] NRC revised its repository licensing regulations to conform to the new EPA standards on April 13, 2009.[105]

DOE estimated in its June 2008 Final Supplemental Environmental Impact Statement (FSEIS) for the Yucca Mountain repository that the maximum mean annual individual dose after 10,000 years would be 2 millirems. That is substantially below the level estimated by the 2002 Final Environmental Impact Statement, which calculated that the peak doses—occurring after 400,000 years—would be about 150 millirems (Volume 1, Chapter 5). The FSEIS attributed the reduction to changes in DOE's computer model and in the assumptions used, noting that "various elements of DOE's modeling approach may be challenged as part of the NRC licensing process."[106]

Alternative Technologies

DOE's Fuel Cycle Research and Development Program conducts "long-term, science-based" research on a wide variety of technologies for improving the management of spent nuclear fuel, as well as implementation of the Administration's integrated waste strategy, according to the FY2016 DOE budget justification. The total FY2016 funding request for this program was $217.8 million, $20.8 million above the FY2015 appropriation. The House Appropriations Committee recommended $175.8 million, eliminating funds for the integrated waste strategy and new facilities for defense and other DOE-managed waste.

A major focus of the Fuel Cycle R&D program is technology related to the reprocessing or "recycling" of spent fuel so that plutonium, uranium, and other long-lived radionuclides could be converted to faster-decaying fission products in special nuclear reactors or particle accelerators. Emplacing waste in deep boreholes, at much greater depths than most proposed repositories, is also being investigated. According to the FY2016 budget justification, "The Department will initiate a field test that will include the drilling of a characterization borehole at a volunteer site that will be selected in the future."

Other long-term disposal alternatives to geologic repositories are disposal below the seabed and transport into space, neither of which is currently being studied by DOE.

Funding and Costs

The Obama Administration's FY2011 budget request called for a complete halt in funding for the Yucca Mountain project and elimination of OCRWM. In line with the request, the FY2011 Continuing Appropriations Act (P.L. 112-10) provided no DOE funding for the program. DOE shut down the Yucca Mountain project at the end of FY2010 and transferred OCRWM's remaining functions to the Office of Nuclear Energy.

President Obama's FY2016 federal budget, submitted to Congress in February 2015, included no funding for Yucca Mountain but requested $30.0 million to conduct research and analysis in support of DOE's January 2013 nuclear waste strategy. The Administration's request included $24.0 million to be appropriated from the Nuclear Waste Fund and a proposal that discretionary appropriations beginning in FY2019 be supplemented by mandatory appropriations. In its FY2015 Energy and Water Development appropriations bill, the House Appropriations Committee provided no funding for DOE's integrated management program but $150.0 million to DOE and $25.0 million for NRC to continue working on the Yucca Mountain license application.

Although nuclear utilities had paid fees to the Nuclear Waste Fund to cover the disposal costs of civilian nuclear spent fuel (until halted by a court order in May 2014), DOE cannot spend the money in the fund until it is appropriated by Congress. At the end of FY2014, the Waste Fund balance stood at $32.9 billion, according to the Treasury Department.[107] Before the Obama Administration halted the Yucca Mountain project, $7.41 billion had been disbursed from the Waste Fund, according to DOE's program summary report.[108] DOE's most recent update of its *Analysis of the Total System Life Cycle Cost of the Civilian Radioactive Waste Management Program* was released on August 5, 2008.[109] According to that estimate, the Yucca Mountain program as then planned would cost $96.2 billion in 2007 dollars from the beginning of the program in 1983 to repository closure in 2133.

Low-Level Radioactive Waste

Current Policy

Selecting disposal sites for low-level radioactive waste, which generally consists of low concentrations of relatively short-lived radionuclides, is authorized to be conducted by states under the 1980 Low-Level Radioactive Waste Policy Act and 1985 amendments. Most states have joined congressionally approved interstate compacts to handle low-level waste disposal. Under the 1985 amendments, the nation's three (at that time) operating commercial low-level waste disposal facilities could start refusing to accept waste from outside their regional interstate compacts after the end of 1992. One of the three sites, near Beatty, NV, closed. The remaining two—at Barnwell, SC, and Hanford, WA—are using their congressionally granted authority to prohibit waste from outside their regional compacts. Another site, in Utah, has since become available nationwide for most Class A low-level waste, but class B and C waste generally must be stored at the sites where it is generated.

The startup of a new disposal facility for Class A, B, and C low-level waste near Andrews, TX, in 2012 may alleviate the class B and C storage problem. Although the facility is intended to serve primarily Texas and Vermont, up to 30% of its 2.3 million cubic feet of disposal capacity may be allocated to waste from other states.[110] The Texas site received its first shipment of waste, from a company in Vermont, on April 27, 2012.[111]

Legislation providing congressional consent to the Texas compact, which originally also included Maine as well as Vermont, was signed by President Clinton September 20, 1998 (P.L. 105-236).

However, on October 22, 1998, a proposed disposal site near Sierra Blanca, TX, was rejected by the Texas Natural Resource Conservation Commission, and Maine subsequently withdrew. Texas Governor Rick Perry signed legislation June 20, 2003, authorizing the Texas Commission on Environmental Quality (TCEQ) to license adjoining disposal facilities for commercial and federally generated low-level waste. Pursuant to that statute, an application to build the Andrews County disposal facility was filed August 2, 2004, by Waste Control Specialists LLC. TCEQ voted January 14, 2009, to issue the license after the necessary land and mineral rights had been acquired and approved construction of the facility January 7, 2011.[112]

The disposal facility at Barnwell, SC, is currently accepting all Class A, B, and C low-level waste from the Atlantic Compact (formerly the Northeast Compact), in which South Carolina joined original members Connecticut and

New Jersey on July 1, 2000. Under the compact, South Carolina can limit the use of the Barnwell facility to the three compact members, and a state law enacted in June 2000 phased out acceptance of non-compact waste through June 30, 2008. The Barnwell facility previously had stopped accepting waste from outside the Southeast Compact at the end of June 1994. The Southeast Compact Commission in May 1995 twice rejected a South Carolina proposal to open the Barnwell site to waste generators outside the Southeast and to bar access to North Carolina until that state opened a new regional disposal facility, as required by the compact. The rejection of those proposals led the South Carolina General Assembly to vote in 1995 to withdraw from the Southeast Compact and begin accepting waste at Barnwell from all states but North Carolina. North Carolina withdrew from the Southeast Compact July 26, 1999. The U.S. Supreme Court ruled on June 1, 2010, that the withdrawal did not subject North Carolina to sanctions under the compact.[113]

The only other existing disposal facility for all three major classes of low-level waste is at Hanford, WA. Controlled by the Northwest Compact, the Hanford site will continue taking waste from the neighboring Rocky Mountain Compact under a contract. Since the South Carolina facility closed to out-of-region waste, and pending planned imports by the Texas compact, the 34 states and the District of Columbia that are outside the Northwest, Rocky Mountain, Atlantic, and Texas compacts have had no disposal site for Class B and C low-level waste.

Regulatory Requirements

Licensing of commercial low-level waste facilities is carried out under the Atomic Energy Act by NRC or by "agreement states" with regulatory programs approved by NRC. NRC regulations governing low-level waste licenses[114] must conform to general environmental protection standards and radiation protection guidelines issued by EPA. Transportation of low-level waste is jointly regulated by NRC and the Department of Transportation.

NRC proposed a significant modification of its low-level waste disposal regulations on March 26, 2015.[115] Among the proposed changes are a requirement that technical analyses for disposal sites include a 1,000-year compliance period for the general public and a dose limit for inadvertent intruders. New analyses would be required for 10,000-year and post-10,000-year periods, along with annual dose minimization targets for the first 10,000 years. NRC's current low-level waste regulations were adopted in 1982.

CONCLUDING DISCUSSION

Disposal of radioactive waste will be a key issue in the continuing nuclear power debate. Without a national disposal system, spent fuel from nuclear power plants must be stored on-site indefinitely. This situation may raise public concern near proposed reactor sites, particularly at sites without existing reactors where spent nuclear fuel is already stored. Concern about spent fuel storage safety has been heightened by the March 2011 disaster at Japan's Fukushima Daiichi nuclear plant.

Under current law, the federal government's nuclear waste disposal policy is focused on the Yucca Mountain site. However, President Obama's moves to terminate the Yucca Mountain project and develop a new waste strategy through the Blue Ribbon Commission on America's Nuclear Future have brought most activities in the DOE waste program to a halt. Congress is continuing to debate the project's termination, particularly through the appropriations process. The NRC staff's finding in October 2014 that the Yucca Mountain would meet NRC standards after the repository was filled and sealed has intensified criticism of the Administration's nuclear waste policy.

Because of their waste-disposal contracts with DOE, owners of existing reactors are likely to continue seeking damages from the federal government if disposal delays continue. For example, DOE's 2004 settlement with the nation's largest nuclear operator, Exelon, could require payments of up to $600 million from the federal judgment fund. DOE estimates that payments could rise above $20 billion if the federal government cannot begin taking waste from reactor sites before 2020, as previously planned. The nuclear industry has predicted that future damages could rise by tens of billions of dollars if the federal disposal program fails altogether.

Lack of a nuclear waste disposal system could also affect the licensing of proposed new nuclear plants, both because of NRC licensing guidelines and various state laws.[116] In addition, further repository delays could force DOE to miss compliance deadlines for defense waste disposal.

Problems being created by nuclear waste disposal delays were addressed by the Blue Ribbon Commission in its final report, issued in January 2012. Major options include centralized interim storage, continued storage at existing nuclear sites, reprocessing and waste treatment technology, development of alternative repository sites, or a combination. The commission recommended that a congressionally chartered corporation be established to undertake a negotiated process for siting new waste storage and disposal facilities. However, given the delays resulting from the ongoing shutdown of the nuclear waste program, longer on-site storage is almost a certainty under any option. Any of the options would also face intense controversy, especially among states and regions that might be potential hosts for future waste facilities. As a result, substantial debate would be expected over any proposals to change the Nuclear Waste Policy Act, including those of the Blue Ribbon Commission.

SELECTED LEGISLATION

H.R. 1364 (Titus)/S. 691 (Reid)

Nuclear Waste Informed Consent Act. Prohibits NRC from authorizing construction of a nuclear waste repository unless the Secretary of Energy has reached an agreement with the host state and with affected units of local government and Indian tribes. House bill introduced March 13, 2015; referred to Committee on Energy and Commerce. Senate bill introduced March 10, 2015; referred to Committee on Environment and Public Works.

H.R. 2028 (Simpson)

Energy and Water Development Appropriations Bill, 2016. Includes funding for licensing the Yucca Mountain nuclear waste repository and nuclear waste R&D. Draft bill approved by the Subcommittee on Energy and Water Development of the Committee on Appropriations on April 15, 2015, and approved by the full committee April 22, 2015. Reported as an original measure by the Committee on Appropriations April 24, 2015 (H.Rept. 114-91).

S. 854 (Alexander)

Nuclear Waste Administration Act of 2015. Establishes an independent Nuclear Waste Administration (NWA) to develop nuclear waste storage and disposal facilities. Siting of such facilities would require the consent of the affected state, local, and tribal governments. NWA would be required to prepare a mission plan to open a pilot storage facility by the end of 2021 for nuclear waste from shutdown reactors and other emergency deliveries (called "priority waste"). A storage facility for waste from operating reactors or other "nonpriority waste" would open by the end of 2025, and a permanent repository by the end of 2048. The current disposal limit of 70,000 metric tons for the nation's first permanent repository would be repealed. Nuclear waste fees collected after enactment of the bill would be held in a newly established Working Capital Fund. The Nuclear Waste Administration could immediately draw from that fund any amounts needed to carry out S. 854, unless limited by annual appropriations or authorizations. Introduced March 24, 2015; referred to Committee on Energy and Natural Resources.

S. 944 (Boxer)

Safe and Secure Decommissioning Act of 2015. Requires NRC to maintain full safety and security requirements at permanently closed reactors until all their spent fuel was moved to dry storage. Introduced April 15, 2015; referred to Committee on Environment and Public Works.

S. 945 (Markey)

Dry Cask Storage Act of 2015. Requires nuclear power plants to develop NRC-approved plans for removing spent fuel from storage pools. Within seven years after such plans had been submitted, spent fuel would have to be transferred to dry storage facilities. After the seven-year period, additional spent fuel would have to be transferred to dry casks within a year after it had been determined to be sufficiently cool. Emergency planning zones would have to be expanded from 10 to 50 miles in radius around any reactor determined by NRC to be out of compliance with its spent fuel transfer plan. The emergency zone for a decommissioned reactor could not be reduced below a 10-mile radius until all its spent fuel had been placed in dry storage. NRC would be authorized to use interest earned by the Nuclear Waste Fund to provide grants to nuclear power plants to transfer spent fuel to dry storage. Introduced April 15, 2015; referred to Committee on Environment and Public Works.

For Additional Reading

Blue Ribbon Commission on America's Nuclear Future. *Report to the Secretary of Energy*, January 2012, http://brc.gov/sites/ default/ files/documents/ brc_finalreport_jan2012.pdf.

Commissioned Papers. 2010-2011. Reports on current nuclear waste issues. http://cybercemetery.unt.edu/archive/brc/20120620214809/ http://brc.gov/index.php?q=library/ documents/commissioned-papers.

Harvard University. John F. Kennedy School of Government. Belfer Center for Science and International Affairs. *The Economics of Reprocessing vs. Direct Disposal of Spent Nuclear Fuel.* DE-FG26-99FT4028. December 2003.

Nuclear Waste Technical Review Board.

Experience Gained from Programs to Manage High-Level Radioactive Waste and Spent Nuclear Fuel in the United States and Other Countries. April 2011. http://www.nwtrb.gov/ reports/reports.html.

Survey of National Programs for Managing High-Level Radioactive Waste and Spent Nuclear Fuel. October 2009. http://www.nwtrb.gov/reports/reports.html.

RAND Corporation. *Managing Spent Nuclear Fuel: Strategy Alternatives and Policy Implications.* 2010. 71 p. http://www.rand.org/content/dam/rand/pubs/monographs/2010/ RAND_MG970.pdf.

University of Illinois. Program in Arms Control, Disarmament, and International Security. *'Plan D' for Spent Nuclear Fuel.* 2009. http://acdis.illinois.edu/publications/207/publicationPlanDforSpent NuclearFuel.html.

U.S. Department of Energy.

Assessment of Disposal Options for DOE-Managed High-Level Radioactive Waste and Spent Nuclear Fuel, October 2014, http://www.energy.gov/ne/downloads/assessment-disposaloptions-doe-managed-high-level-radioactive-waste-and-spent-nuclear.

Strategy for the Management and Disposal of Used Nuclear Fuel and High-Level Radioactive Waste, January 2013, http://energy.gov/sites/prod/files/ Strategy%20for%20the%20Management%20and%20Disposal%20of%20Used%20Nuclear%20Fuel%20and%20High%20Level%20Radioactive%20Waste.pdf. *Used Fuel Disposition Campaign: Disposal Research and Development Roadmap*, March 2011, http://energy.gov/sites/prod/ files/UFD_Disposal_R%26D_Roadmap_Rev_0.1.pdf.

U.S. Nuclear Regulatory Commission. *Safety Evaluation Report Related to Disposal of High-Level Radioactive Wastes in a Geologic Repository at Yucca Mountain, Nevada, Volume 3: Repository Safety After Permanent Closure*, NUREG-1949, V3, ML14288A121, October 16, 2014, 781 p.

Walker, J. Samuel. *The Road to Yucca Mountain: The Development of Radioactive Waste Policy in the United States.* University of California Press. 2009. 228 p.

End Notes

[1] NRC, "NRC Staff Issues Volume 3 of Yucca Mountain Safety Evaluation Report," news release 14-069, October 16, 2014, http://www.nrc.gov/reading-rm/doc-collections/nuregs/staff/sr1949/v3/.

[2] NRC, "NRC Publishes Final Two Volumes of Yucca Mountain Safety Evaluation," news release 15-005, January 29. 2015, http://www.nrc.gov/reading-rm/doc-collections/news/2015/.

[3] NRC, "NRC Staff to Prepare Supplement to Yucca Mountain Environmental Impact Statement," news release 15-016, March 12, 2015, http://www.nrc.gov/reading-rm/doc-collections/news/2015/.

[4] Hiruo, Elaine, and Steven Dolley, "NRC Says Staff Can Finish Yucca Supplemental EIS in 12-15 Months," NuclearFuel, March 16, 2015.

[5] Energy Secretary Ernest Moniz, "A Look Back on the Blue Ribbon Commission on America's Nuclear Future," remarks to Bipartisan Policy Center, March 24, 2015, http://www.energy.gov/articles/secretary-monizs-remarks-lookback-blue-ribbon-commission-america-s-nuclear-future.

[6] DOE, Assessment of Disposal Options for DOE-Managed High-Level Radioactive Waste and Spent Nuclear Fuel, October 2014, p. iii.

[7] DOE, Strategy for the Management and Disposal of Used Nuclear Fuel and High-Level Waste, January 2013, http://energy.gov/sites/prod/files/ Strategy%20for%20the%20Management%20and%20Disposal%20of%20Used%20Nuclear%20Fuel%20and%20High%20Level%20Radioactive%20Waste.pdf.

[8] House Committee on Energy and Commerce, "Committee Leaders Respond to DOE's Nuclear Waste Delay," March 24, 2015, http://energycommerce.house.gov/press-release/committee-leaders-respond-doe%E2%80%99snuclear-waste-delay.

[9] Valhi, Inc., "Valhi's Waste Control Specialists Subsidiary to Apply for License to Store Used Nuclear Fuel," February 7, 2015, http://www.valhi.net/phoenix.zhtml?c=103380&p=irol-newsArticle&ID=2014598.

[10] Beattie, Jeff, "Waste Control Specialists Sets 2020 Date to Open Spent Fuel Storage Facility," IHS The Energy Daily, February 10, 2015, p. 1; Hiruo, Elaine, "Texas Company Seeks License for Spent Fuel Storage," Nucleonics Week, February 12, 2015, p. 1.

[11] DOE, "FY 2016 Congressional Budget," Vol. 3, p. 466, http://energy.gov/cfo/downloads/fy-2016-budgetjustification.

[12] Hiruo, Elaine, "DOE Implements Court-Ordered Suspension of Nuclear Waste Fee," NuclearFuel, May 26, 2014.

[13] See CRS Legal Sidebar WSLG734, Court Neither Razzled Nor Dazzled by DOE's Failure to Assess Nuclear Waste Fund Fee, by Todd Garvey.

[14] NRC, "NRC Approves Final Rule on Spent Fuel Storage and Ends Suspension of Final Licensing Actions for Nuclear Plants and Renewals," news release, August 26, 2014, http://pbadupws.nrc.gov/docs/ML1423/ ML14238A326.pdf.

[15] U.S. Circuit Court of Appeals for the District of Columbia Circuit, State of New York, et al. v. Nuclear Regulatory Commission, No. 11-1045, Decided June 8, 2012, http://www.cadc.uscourts.gov/internet/opinions.nsf/ 57ACA94A8FFAD8AF85257A1700502AA4/$file/11-1045-1377720.pdf.

[16] Nuclear Regulatory Commission, "NRC to Issue Orders, Information Request as Part of Implementing FukushimaRelated Recommendations," press release, March 9, 2012, http://www.nrc.gov/reading-rm/doc-collections/news/2012/ 12-023.pdf.

[17] NRC, "Low-Level Radioactive Waste Disposal; Proposed Rule," 80 Federal Register 16082, March 26, 2015.

[18] The term "spent nuclear fuel" is defined by NWPA as "fuel that has been withdrawn from a nuclear reactor following irradiation, the constituent elements of which have not been separated by reprocessing." The nuclear industry refers to this material as "used fuel," because it contains potentially reusable uranium and plutonium.

[19] U.S. Department of Energy's Motion to Withdraw, NRC Atomic Safety and Licensing Board, Docket No. 63-0001, March 3, 2010, http://www.energy.gov/news/documents/DOE_Motion_to_Withdraw.pdf.

[20] National Association of Regulatory Utility Commissioners, "NARUC Seeks Party Status at NRC, Says Yucca Review Must Continue," press release, March 16, 2010, http://www.naruc.org/News/default.cfm?pr=191&pdf=.

[21] U.S. Nuclear Regulatory Commission, Atomic Safety and Licensing Board, Docket No. 63-001-HLW, Memorandum and Order, June 29, 2010.

[22] Nuclear Regulatory Commission, "In the Matter of U.S. Department of Energy (High-Level Waste Repository)," CLI-11-07, September 9, 2011, http://www.nrc.gov/reading-rm/doc-collections/commission/orders/2011/2011- 07cli.pdf.

[23] U.S. Court of Appeals for the District of Columbia Circuit, In re: Aiken County et al., No. 11-1271, writ of mandamus, August 13, 2013, http://www.cadc.uscourts. gov/internet/opinions.nsf/ BAE0CF34F762EBD985257BC6004DEB18/$file/11-1271-1451347.pdf.

[24] Nuclear Regulatory Commission, "NRC Directs Staff to Complete Yucca Mountain Safety Evaluation Report," news release No. 13-097, November 18, 2013, http://pbadupws.nrc.gov/docs/ML1332/ML13322B228.pdf.

[25] NRC, "NRC Staff Issues Volume 3 of Yucca Mountain Safety Evaluation Report," news release 14-069, October 16, 2014, http://www.nrc.gov/reading-rm/doc-collections/nuregs/staff/sr1949/v3/.

[26] NRC, "NRC Publishes Final Two Volumes of Yucca Mountain Safety Evaluation," news release 15-005, January 29. 2015, http://www.nrc.gov/reading-rm/doc-collections/news/2015/.

[27] NRC, "NRC Staff to Prepare Supplement to Yucca Mountain Environmental Impact Statement," news release 15- 016, March 12, 2015, http://www.nrc.gov/reading-rm/doc-collections/news/2015/.

[28] Hiruo, Elaine, and Steven Dolley, "NRC Says Staff Can Finish Yucca Supplemental EIS in 12-15 Months," NuclearFuel, March 16, 2015.

[29] Blue Ribbon Commission on America's Nuclear Future, Report to the Secretary of Energy, January 2012, http://brc.gov/sites/default/files/documents/brc_finalreport_jan2012.pdf (BRC Final Report).

[30] Blue Ribbon Commission on America's Nuclear Future, Commissioned Papers, http://cybercemetery.unt.edu/ archive/brc/20120620214809/http://brc.gov/index.php?q=library/documents/commissioned-papers.

[31] DOE, Strategy for the Management and Disposal of Used Nuclear Fuel and High-Level Waste, op. cit.

[32] Nuclear Energy Institute, "Nuclear Waste Management: Disposal," October 28, 2014, http://www.nei.org/IssuesPolicy/Nuclear-Waste-Management/Disposal.

[33] National Research Council, Board on Radioactive Waste Management, Rethinking High-Level Radioactive Waste Disposal: A Position Statement of the Board on Radioactive Waste Management (1990), p. 2.

[34] Senator Bingaman introduced the approval resolution in the Senate April 9, 2002 (S.J.Res. 34), and Representative Barton introduced it in the House April 11, 2002 (H.J.Res. 87). The Subcommittee on Energy and Air Quality of the House Committee on Energy and Commerce approved H.J.Res. 87 on April 23 by a 24-2 vote, and the full committee approved the measure two days later, 41-6 (H.Rept. 107-425). The resolution was passed by the House May 8, 2002, by a vote of 306-117. The Senate Committee on Energy and Natural Resources approved S.J.Res. 34 by a 13-10 vote June 5, 2002 (S.Rept. 107-159). Following a 60-39 vote to consider S.J.Res. 34, the Senate passed H.J.Res. 87 by voice vote July 9, 2002.

[35] Waste Control Specialists LLC, "Historic Texas Compact Disposal Facility Ready for Business," http://www.wcstexas.com.

[36] Waste Control Specialists LLC, "Waste Control Specialists Commends Passage of Legislation," press release, May 31, 2011, http://www.wcstexas.com/PDF_downloads/WCSAnnounceslegislation.pdf?nxd_id=98546.

[37] The Standard Contract for Disposal of Spent Nuclear Fuel and/or High-Level Radioactive Waste can be found at 10 CFR 961.11.

[38] DOE, Agency Financial Report Fiscal Year 2014, DOE/CF-0106, November 14, 2014, p. 76, http://energy.gov/cfo/ downloads/fy-2014-doe-agency-financial-report.

[39] DOE, Agency Financial Report Fiscal Year 2014, op. cit.

[40] NARUC, "State Regulators Go to Court with DOE over Nuclear Waste Fees, news release, April 2, 2010, http://www.naruc.org/News/default.cfm?pr=193; Nuclear Energy Institute et al. v. U.S. DOE, Joint Petition for Review, U.S. Court of Appeals for the District of Columbia Circuit, April 5, 2010.

[41] Jeff Beattie, "NARUC, Utilities Sue DOE Over Nuke Waste Fee," Energy Daily, April 6, 2010, p. 1.

[42] Secretary of Energy Steven Chu, "Secretarial Determination of the Adequacy of the Nuclear Waste Fund Fee," November 1, 2010, http://energy.gov/sites/prod/files/gcprod/documents/Secretarial_Determination_WasteFee.pdf.

[43] U.S. Court of Appeals for the District of Columbia Circuit, National Association of Regulatory Utility Commissioners v. United States Department of Energy, No. 11-1066, decided June 1, 2012, http://www.cadc.uscourts.gov/internet/opinions.nsf/ 4B11622F4FF7 5FEC85257A 100050A681/$file/11-1066- 1376508.pdf.

[44] DOE, "Nuclear Waste Fund Fee Adequacy Report," January 2013, http://energy.gov/sites/prod/files/
January%2016%202013%20Secretarial%20Determination%20of%20the%20Adequacy%20of%20the%20Nuclear%20 Waste%20Fund%20Fee_0.pdf.

[45] See CRS Legal Sidebar WSLG734, Court Neither Razzled Nor Dazzled by DOE's Failure to Assess Nuclear Waste Fund Fee, by Todd Garvey.

[46] Hiruo, Elaine, "DOE Implements Court-Ordered Suspension of Nuclear Waste Fee," NuclearFuel, May 26, 2014.

[47] Statement of Peter B. Lyons, Assistant Secretary for Nuclear Energy, U.S. Department of Energy, before the Committee on Energy and Commerce, Subcommittee on Environment and the Economy, June 1, 2011.

[48] DOE Motion to Withdraw, op. cit.

[49] Nicole E. Matthews, "DOE Withdraws Application for Yucca Nuke Dump," Fox5Vegas.com, March 3, 2010, http://www.fox5vegas.com/news/22734591/detail.html.

[50] Motion for the State of Nevada for Leave to Intervene as Intervenor-Respondent, U.S. Court of Appeals for the Fourth Circuit, Case No. 10-1229, March 19, 2010, http://www.state.nv.us/nucwaste/licensing/nv100319motion3.pdf.

[51] National Association of Regulatory Utility Commissioners, "NARUC Seeks Party Status at NRC, Says Yucca Review Must Continue," press release, March 16, 2010, http://www.naruc.org/News/default.cfm?pr=191&pdf=.

[52] U.S. Nuclear Regulatory Commission, Atomic Safety and Licensing Board, Docket No. 63-001-HLW, Memorandum and Order, June 29, 2010.

[53] U.S. Department of Energy's Brief in Support of Review and Reversal of the Board's Ruling on the Motion to Withdraw, Docket No. 63-001-HLW, July 9, 2010.

[54] Nuclear Regulatory Commission, Memorandum and Order, CLI-11-07, September 9, 2011, http://www.nrc.gov/ reading-rm/doc-collections/commission/orders/2011/2011-07cli.pdf.

[55] U.S. Circuit Court of Appeals for the District of Columbia Circuit, USCA Case #11-1271, Yucca Mountain Reply Brief of Petitioners Mandamus Action, February 13, 2012, http://www.naruc.org/policy.cfm?c=filings.

[56] Freebairn, William, "NRC Says Lack of Funding Supports Decision to Stop Yucca Review," NuclearFuel, January 21, 2013.

[57] U.S. Court of Appeals for the District of Columbia Circuit, In re: Aiken County et al., No. 11-1271, writ of mandamus, August 13, 2013, http://www.cadc.uscourts.gov /internet/opinions.nsf/BAE0CF34F762EBD985257BC6004DEB18/$file/11-1271-1451347.pdf.

[58] Hiruo, Elaine, and Steven Dolley, "NRC Says Staff Can Finish Yucca Supplemental EIS in 12-15 Months," NuclearFuel, March 16, 2015.

[59] U.S. Circuit Court of Appeals for the District of Columbia Circuit, Minnesota v. NRC, 602 F.2d 412 (D.C. Cir. 1979).

[60] NRC, "Waste Confidence Decision," 49 Federal Register 34,658, August 31, 1984.

[61] NRC, "Waste Confidence Decision Review," 55 Federal Register 38,474, September 18, 1990.

[62] NRC, "Waste Confidence Decision Update," 75 Federal Register 81,037, December 23, 2010.

[63] U.S. Circuit Court of Appeals for the District of Columbia Circuit, State of New York, et al. v. Nuclear Regulatory Commission, No. 11-1045, Decided June 8, 2012, http://www.cadc.uscourts.gov/internet/opinions.nsf/ 57ACA94A8FFAD8AF85257A1700502AA4/$file/11-1045-1377720.pdf.

[64] NRC, "NRC Approves Final Rule on Spent Fuel Storage and Ends Suspension of Final Licensing Actions for Nuclear Plants and Renewals," news release, August 26, 2014, http://pbadupws.nrc.gov/docs/ML1423/ ML14238A326.pdf.

[65] NRC, "Continued Storage of Spent Nuclear Fuel," 79 Federal Register 56238, September 19, 2014. Available at NRC, "Continued Storage of Spent Nuclear Fuel," http://www.nrc.gov/waste/spent-fuel-storage/wcd/documents.html.

[66] Southern Alliance for Clean Energy, news release, September 29, 2014, http://www.cleanenergy.org/17-groups-urgenrc-to-halt-licensing-relicensing-of-23-reactors-due-to-failure-to-address-2012-court-ruling/.

[67] State of Vermont, Office of the Attorney General, "Vermont Joins Challenge to Nuclear Regulatory Commission's Rule on Storing Spent Nuclear Fuel," October 27, 2014,

http://ago.vermont.gov/focus/news/vermont-joins-challengeto-nuclear-regulatory-commissions-rule-on-storing-spent-nuclear-fuel.php.

[68] Senate Committee on Energy and Natural Resources, Full Committee Hearing to Consider the Nuclear Waste Administration Act of 2013, July 30, 2013, http://www.energy.senate.gov/public/index.cfm/2013/7/full-committeehearing-to-consider-the-nuclear-waste-administration-act-of-2013.

[69] Senate Committee on Appropriations, "FY15 Subcommittee Reported Bill and Draft Report," July 24, 2014, http://www.appropriations.senate.gov/news/fy-2015-ew-subcommittee-reported-bill-and-draft-report.

[70] Nuclear Regulatory Commission, "Staff Evaluation and Recommendations for Japan Lessons-Learned Tier 3 Issue on Expedited Transfer of Spent Fuel," COMSECY-13-0030, November 12, 2013, http://www.nrc.gov/reading-rm/doccollections/commission/comm-secy/2013/2013-0030comscy.pdf.

[71] Statutory definitions for "spent nuclear fuel," "high-level radioactive waste," and "low-level radioactive waste" can be found in Section 2 of the Nuclear Waste Policy Act of 1982 (42 U.S.C. 10101). "Transuranic waste" is defined in Section 11ee. of the Atomic Energy Act (42 U.S.C. 2014); Section 11e.(2) of the act includes uranium mill tailings in the definition of "byproduct material." "Mixed waste" consists of chemically hazardous waste as defined by EPA regulations (40 CFR Part 261, Subparts C and D) that contains radioactive materials as defined by the Atomic Energy Act.

[72] DOE, "Waste Isolation Pilot Plant Recovery," October 28, 2014, http://www.wipp.energy.gov/wipprecovery/ recovery.html

[73] Nuclear Regulatory Commission, "Staff Evaluation and Recommendations for Japan Lessons-Learned Tier 3 Issue on Expedited Transfer of Spent Fuel," op. cit., Enclosure 1, p. 77.

[74] Gutherman Technical Services, 2012 Used Fuel Data, January 30, 2013. Adjusted for four sites closed during 2013 and 2014. Includes General Electric's spent fuel storage facility at Morris, IL, located adjacent to the Dresden nuclear plant. Also, the Hope Creek and Salem nuclear plants in New Jersey are counted as a single site.

[75] Gutherman Technical Services, 2013 Used Fuel Data, January 20, 2014.

[76] DOE Office of Civilian Radioactive Waste Management, OCRWM Annual Report to Congress, Fiscal Year 2002, DOE/RW-0560, October 2003, Appendix C.

[77] DOE Office of Civilian Radioactive Waste Management, Draft Supplemental Environmental Impact Statement for a Geologic Repository for the Disposal of Spent Nuclear Fuel and High-Level Radioactive Waste at Yucca Mountain, Nye County, Nevada, Summary, DOE/EIS-0250F-S1D, October 2007, p. S-47.

[78] Gutherman, 2014. Excludes DOE facilities and the unconstructed Private Fuel Storage facility in Utah. In addition, GE operates an independent pool storage facility near Morris, IL.

[79] National Academy of Sciences, Safety and Security of Commercial Spent Nuclear Fuel Storage: Public Report, released April 6, 2005, p. 2.

[80] Nuclear Regulatory Commission, "Staff Evaluation and Recommendations for Japan Lessons-Learned Tier 3 Issue on Expedited Transfer of Spent Fuel," op. cit.

[81] U.S. Nuclear Regulatory Commission, Near-Term Task Force Review of Insights from the Fukushima Dai-ichi Accident, Recommendations for Enhancing Reactor Safety in the 21st Century, p. 46, http://pbadupws.nrc.gov/docs/ ML1118/ML111861807.pdf.

[82] Nuclear Regulatory Commission, "NRC to Issue Orders, Information Request as Part of Implementing FukushimaRelated Recommendations," press release, March 9, 2012, http://www.nrc.gov/reading-rm/doc-collections/news/2012/ 12-023.pdf.

[83] U.S. Department of Energy, Management Information Manifest System, http://mims.doe .gov/GeneratorData.aspx. Most recent year reported.

[84] BRC Final Report, op. cit.

[85] DOE, Strategy for the Management and Disposal of Used Nuclear Fuel and High-Level Waste, op. cit.

[86] Spencer, Jack, "Nuclear Waste Management: Minimum Requirements for Reforms and Legislation," Heritage Foundation, March 28, 2013, http://www.heritage.org/research/ reports/2013/03/nuclear-waste-management-minimumrequirements-for-reforms-and-legislation.

[87] DOE, Assessment of Disposal Options for DOE-Managed High-Level Radioactive Waste and Spent Nuclear Fuel, October 2014, http://www.energy.gov/ne/downloads/assessment-disposal-options-doe-managed-high-level-radioactivewaste-and-spent-nuclear.

[88] Energy Secretary Ernest Moniz, "A Look Back on the Blue Ribbon Commission on America's Nuclear Future," remarks to Bipartisan Policy Center, March 24, 2015, http://www.energy.gov/articles/secretary-monizs-remarks-lookback-blue-ribbon-commission-america-s-nuclear-future.

[89] Valhi, Inc., "Valhi's Waste Control Specialists Subsidiary to Apply for License to Store Used Nuclear Fuel," February 7, 2015, http://www.valhi.net/phoenix.zhtml?c=103380&p=irol-newsArticle&ID=2014598.

[90] Beattie, Jeff, "Waste Control Specialists Sets 2020 Date to Open Spent Fuel Storage Facility," IHS The Energy Daily, February 10, 2015, p. 1; Hiruo, Elaine, "Texas Company Seeks License for Spent Fuel Storage," Nucleonics Week, February 12, 2015, p. 1.

[9]1 Elaine Hiruo and Herman Wang, "New Mexico Interested in Facilities for Waste Storage, Disposal: Officials," NuclearFuel, March 5, 2012, p. 3.

[92] Maddox, James M., Chairman, Eddy-Lea Energy Alliance, LLC, "Notice of Intent to Submit a License Application for Consolidated Used Nuclear Fuel Storage Facility," February 26, 2013, http://pbadupws.nrc.gov/docs/ML1306/ ML13067A278.pdf.

[93] Housley Carr and Elaine Hiruo, "Group Urges Mississippi to Become Home to Spent Fuel Facilities," NuclearFuel, September 2, 2013.

[94] Bureau of Indian Affairs, Record of Decision for the Construction and Operation of an Independent Spent Fuel Storage Installation (ISFSI) on the Reservation of the Skull Valley Band of Goshute Indians (Band) in Tooele County, Utah, September 7, 2006.

[95] Bureau of Land Management, Record of Decision Addressing Right-of-Way Applications U 76985 and U 76986 to Transport Spent Nuclear Fuel to the Reservation of the Skull Valley Band of Goshute Indians, September 7, 2006.

[96] Winslow, Ben, "Goshutes, PFS Sue Interior," Deseret Morning News, July 18, 2007.

[97] U.S. District Court for the District of Utah, Skull Valley Band of Goshute Indians and Private Fuel Storage v. United States Department of the Interior, Civil Action No. 07-cv-0526-DME-DON, July 26, 2010, http://64.38.12.138/docs/ court/goshute/order072610.pdf.

[98] Palmberg, Robert M., Chairman of the Board, Private Fuel Storage LLC, letter to Nuclear Regulatory Commission, December 20, 2012, http://pbadupws.nrc.gov/docs/ ML1235/ ML12356A063.pdf.

[99] See, for example: NRC, "Analysis of Energy Policy Act of 1992 Issues Related to High-Level Waste Disposal Standards, SECY-93-013, January 25, 1993, attachment p. 4.

[100] National Research Council. Technical Bases for Yucca Mountain Standards. National Academy Press. 1995.

[101] Nuclear Energy Institute v. Environmental Protection Agency, U.S. Court of Appeals for the District of Columbia Circuit, No. 01-1258, July 9, 2004.

[102] Especially high doses at the upper end of the exposure range would raise the mean, or average, more than the median, or the halfway point in the data set.

[103] Office of the Governor, Agency for Nuclear Projects. Comments by the State of Nevada on EPA's Proposed New Radiation Protection Rule for the Yucca Mountain Nuclear Waste Repository. November 2005.

[104] Posted on the EPA website at http://www.epa.gov/radiation/yucca.

[105] Nuclear Regulatory Commission, "Implementation of a Dose Standard After 10,000 Years," 74 Federal Register 10811, March 13, 2009.

[106] FSEIS, p. S-42. Posted on the DOE website at http://www.rw.doe.gov/ ym_ repository/ seis/docs/002_Summary.pdf.

[107] DOE, Office of Inspector General, Department of Energy's Nuclear Waste Fund's Fiscal Year 2014 Financial Statement Audit, OAS-FS-15-03, November 24, 2014, p. 8, http://energy.gov/ig/downloads/audit-report-oas-fs-15-03. Excludes $3.1 billion in unpaid one-time fees for waste generated before enactment of NWPA.

[108] DOE, Office of Civilian Radioactive Waste Management, Office of Program Management, Monthly Summary of Program Financial and Budget Information, as of January 31, 2010, available at http://www.thenwsc.org/ym/ DOE%20Financial%20&%20Budget% 20 Summary%20013110.pdf. The report notes that some figures may not add due to independent rounding.

[109] Available on the DOE website at http://energy.gov/sites/prod/files/gcprod/documents/ FY_2007_TotalSystemLifeCycleCost_Pub2008.pdf.

[110] Waste Control Specialists, "Our Facilities: Compact Waste Facility," http://www.wcstexas. com/facilities/compactwaste-facility/.

[111] Bionomics, Inc., "Bionomics Makes First Shipment to Texas Low Level Waste Site," press release, April 27, 2012, http://www.bionomics-inc.com/documents/Newsletter/First%2 0Shipment%20to%20Texas.pdf.

[112] See the TCEQ website, http://www.tceq.state.tx.us/permitting/radmat/licensing/ wcs_ license_app.html#wcs_status.

[113] Alabama et al. v. North Carolina, S. Ct. (2010), http://www.supremecourt.gov/opinions/ 09pdf/132Orig.pdf.

[114] 10 Code of Federal Regulations Part 61, Licensing Requirements for Land Disposal of Radioactive Waste.

[115] NRC, "Low-Level Radioactive Waste Disposal; Proposed Rule," 80 Federal Register 16082, March 26, 2015.

[116] Lovell, David L., Wisconsin Legislative Council Staff, State Statutes Limiting the Construction of Nuclear Power Plants, October 5, 2006.

In: Challenges Facing the Nuclear Power Industry ISBN: 978-1-63483-618-0
Editor: Renee Olson

Chapter 4

NUCLEAR POWER PLANT SECURITY AND VULNERABILITIES*

Mark Holt and Anthony Andrews

SUMMARY

The physical security of nuclear power plants and their vulnerability to deliberate acts of terrorism was elevated to a national security issue following the attacks of September 11, 2001. Congress subsequently enacted new nuclear plant security requirements and has repeatedly focused attention on regulation and enforcement by the Nuclear Regulatory Commission (NRC). More than a decade after the 9/11 attacks, security at nuclear plants remains an important concern.

The Energy Policy Act of 2005 (EPACT05, P.L. 109-58) imposed specific criteria for NRC to consider in revising the "Design Basis Threat" (DBT), which specifies the maximum severity of potential attacks that a nuclear plant's security force must be capable of repelling. In response to the legislative mandate, NRC revised the DBT (10 C.F.R. Part 73.1) on April 18, 2007. Among other changes, the revisions expanded the assumed capabilities of adversaries to operate as one or more teams and attack from multiple entry points.

To strengthen nuclear plant security inspections, EPACT05 required NRC to conduct "force-onforce" security exercises at nuclear power

* This is an edited, reformatted and augmented version of a Congressional Research Service publication RL34331, prepared for Members and Committees of Congress dated January 3, 2014.

plants at least once every three years. In these exercises, a mock adversary force from outside a nuclear plant attempts to penetrate the plant's vital area and simulate damage to a "target set" of key safety components. During calendar year 2012, NRC conducted 23 force-on-force (FOF) inspections at 22 commercial nuclear plants and one fuel cycle facility. Eleven of those inspections found performance deficiencies: 19 with low significance (green findings), one "greater than green" finding, and three severity level IV (least serious) violations. One exercise resulted in the simulated destruction of or damage to a complete "target set" of vital plant components that were under mock attack. When force-on-force exercises indicate inadequate protection against the DBT, additional security measures must be promptly implemented, according to NRC.

Nuclear power plant vulnerability to deliberate aircraft crashes has been a continuing issue. After much consideration, NRC published final rules on June 12, 2009, to require all new nuclear power plants to incorporate design features that would ensure that, in the event of a crash by a large commercial aircraft, the reactor core would remain cooled or the reactor containment would remain intact, and radioactive releases would not occur from spent fuel storage pools.

NRC rejected proposals that existing reactors also be required to protect against aircraft crashes, such as by adding large external steel barriers, deciding that other mitigation measures already required by NRC for all reactors were sufficient. In 2002, NRC ordered all nuclear power plants to develop strategies to mitigate the effects of large fires and explosions that could result from aircraft crashes or other causes. NRC published a broad final rule on nuclear reactor security March 27, 2009, including fire mitigation strategies and requirements that reactors establish procedures for responding to specific aircraft threats.

Other ongoing nuclear plant security issues include the vulnerability of spent fuel pools, which hold highly radioactive nuclear fuel after its removal from the reactor, standards for nuclear plant security personnel, and nuclear plant emergency planning. NRC's March 2009 security regulations addressed some of those concerns and included a number of other security enhancements.

OVERVIEW OF REACTOR SECURITY

Physical security at nuclear power plants involves the threat of radiological sabotage—a deliberate act against a plant that could directly or indirectly endanger public health and safety through exposure to radiation. The Nuclear Regulatory Commission (NRC) establishes security requirements at

U.S. commercial nuclear power plants based on its assessment of plant vulnerabilities to, and the consequences of, potential attacks. The stringency of NRC's security requirements and its enforcement program have been a significant congressional issue, especially since the September 11, 2001, terrorist attacks on the United States.

While NRC establishes security requirements within the boundaries of commercial nuclear sites, the Department of Homeland Security (DHS) has broad responsibility for coordinating government-wide efforts to prevent and respond to terrorist attacks, including attacks on nuclear power plants. DHS works with NRC and other agencies to protect nuclear facilities and other critical infrastructure.[1]

Nuclear plant security measures are designed to protect three primary areas of vulnerability: controls on the nuclear chain reaction, cooling systems that prevent hot nuclear fuel from melting even after the chain reaction has stopped, and storage facilities for highly radioactive spent nuclear fuel. U.S. plants are designed and built to prevent dispersal of radioactivity, in the event of an accident, by surrounding the reactor in a steel-reinforced concrete containment structure. However, as the March 2011 Fukushima disaster in Japan demonstrated, reactor containments cannot completely block radioactive releases under the most severe circumstances, such as when cooling systems are disabled for several hours or longer.

NRC requires commercial nuclear power plants to have a series of physical barriers and a trained security force, under regulations already in place prior to the 9/11 attacks (10 C.F.R. 73—Physical Protection of Plants and Materials). The plant sites are divided into three zones: an "owner-controlled" buffer region, a "protected area," and a "vital area." Access to the protected area is restricted to a portion of plant employees and monitored visitors, with stringent access barriers. The vital area is further restricted, with additional barriers and access requirements. The security force must comply with NRC requirements on pre-hiring investigations and training.[2]

A fundamental concept in NRC's physical security requirements is the design basis threat (DBT), which establishes the severity of the potential attacks that a nuclear plant's security force must be capable of repelling. The DBT includes such characteristics as the number of attackers, their training, and the weapons and tactics they could use. Specific details are classified. Critics of nuclear plant security have contended that the DBT should be strengthened to account for potentially larger and more sophisticated terrorist attacks.

Reactor vulnerability to deliberate aircraft crashes has also been a major concern since 9/11. Most existing nuclear power plants were not specifically designed to withstand crashes from large jetliners, although analyses differ as to the damage that could result. NRC has determined that commercial aircraft crashes are beyond the DBT but published regulations in June 2009 to require that new reactor designs be able to withstand such crashes without releasing radioactivity. Nuclear power critics have called for retrofits of existing reactors as well.

Since the 9/11 attacks, NRC and Congress have taken action to increase nuclear power plant security. NRC issued a series of security measures beginning in 2002, including a strengthening of the DBT and establishing the Office of Nuclear Security and Incident Response (NSIR). The office centralizes security oversight of all NRC-regulated facilities, coordinates with law enforcement and intelligence agencies, and handles emergency planning activities. In 2004, NRC implemented a program to conduct "force on force" security exercises overseen by NSIR at each nuclear power plant at least every three years. The Energy Policy Act of 2005 (P.L. 109-58) required NRC to further strengthen the DBT, codified the force-on-force program, and established a variety of additional nuclear plant security measures. In March 2009, NRC published a series of security regulations that require power plants to prepare cybersecurity plans, develop strategies for dealing with the effects of aircraft crashes, strengthen access controls, improve training for security personnel, and implement other new security measures.

DESIGN BASIS THREAT

The design basis threat describes general characteristics of adversaries that nuclear plants and nuclear fuel cycle facilities must defend against to prevent radiological sabotage and theft of strategic special nuclear material. NRC licensees use the DBT as the basis for implementing defensive strategies at specific nuclear plant sites through security plans, safeguards contingency plans, and guard training and qualification plans.

General requirements for the DBT are prescribed in NRC regulations,[3] while specific attributes of potential attackers, such as their weapons and ammunition, are contained in classified adversary characteristics documents (ACDs).

Fundamental policies on nuclear plant security threats date back to the Cold War. In 1967, the Atomic Energy Commission (AEC) instituted a rule that nuclear plants are not required to protect against an attack directed by an "enemy of the United States."[4] That so-called "Enemy of the State Rule" specifies that nuclear power plants are

> not required to provide for design features or other measures for the specific purpose of protection against the effects of (a) attacks and destructive acts, including sabotage, directed against the facility by an enemy of the United States, whether a foreign government or other person, or (b) use or deployment of weapons incident to U.S. defense activities.[5]

The Nuclear Regulatory Commission (NRC), the AEC's successor regulatory agency, says that the rule "was primarily intended to make clear that privately-owned nuclear facilities were not responsible for defending against attacks that typically could only be carried out by foreign military organizations."[6] NRC's initial DBT, established in the late 1970s, was intended to be consistent with the enemy of the state rule, which remains in effect.

However, the 9/11 attacks drew greater attention to the potential severity of credible terrorist threats. Following the attacks, NRC evaluated the extent to which nuclear plant security forces should be able to defend against such threats, and ordered a strengthening of the DBT, along with other security measures, on April 29, 2003. That order changed the DBT to "represent the largest reasonable threat against which a regulated private guard force should be expected to defend under existing law," according to the NRC announcement.[7]

In the Energy Policy Act of 2005 (EPACT05), Congress imposed a statutory requirement on the NRC to initiate rulemaking for revising the design basis threat.[8] EPACT05 required NRC to consider 12 factors in revising the DBT, such as an assessment of various terrorist threats, the potential use of substantial explosive devices and modern weapons, attacks by persons with sophisticated knowledge of facility operations, and attacks on spent fuel shipments.

NRC approved its final rule amending the DBT (10 C.F.R. Part 73.1) on January 29, 2007, effective April 18, 2007.[9] Although specific details of the revised DBT were not released to the public, in general the final rule

- clarifies that physical protection systems are required to protect against diversion and theft of fissile material;

- expands the assumed capabilities of adversaries to operate as one or more teams and attack from multiple entry points;
- assumes that adversaries are willing to kill or be killed and are knowledgeable about specific target selection;
- expands the scope of vehicles that licensees must defend against to include water vehicles and land vehicles beyond the four-wheel-drive type;
- revises the threat posed by an insider to be more flexible in scope; and
- adds a new mode of attack from adversaries coordinating a vehicle bomb assault with another external assault.

The DBT final rule excluded aircraft attacks as beyond the reasonable responsibility of a private security force, a decision that raised considerable controversy. In approving the rule, NRC rejected a petition from the Union of Concerned Scientists to require that nuclear plants be surrounded by aircraft barriers made of steel beams and cables (the so-called "beamhenge" concept). Critics of NRC's final rule charged that deliberate aircraft crashes were a highly plausible mode of attack, given the events of 9/11. However, NRC contended that power plants were already required to mitigate the effects of aircraft crashes and that "active protection against airborne threats is addressed by other federal organizations, including the military."[10] Additional NRC action on aircraft threats is discussed in the next section.

NRC Commissioners in January 2009 rejected a proposal by the NRC staff to strengthen the classified portion of the DBT to include additional capabilities by potential attackers, according to news reports. The staff proposal lost in a 2-2 vote, with one commissioner position vacant. Critics contend that the DBT excludes major types of weapons used by terrorists, such as rocket-propelled grenades, and is generally not based on the maximum credible threat identified by the intelligence community.[11]

Critics of NRC's security regulations also have pointed out that licensees are required to employ only a minimum of 10 security personnel on duty per plant, which they argue is not enough for the job.[12] Nuclear spokespersons have responded that the actual security force for the nation's 65 nuclear plant sites numbers more than 5,000, an average of about 75 per site (covering multiple shifts). The industry also points out that nuclear plants all have integrated communications and emergency response plans that include local, state, and federal security forces. The integrated response by outside security forces is intended to handle attacks that might overwhelm an individual plant's security force.[13]

LARGE AIRCRAFT CRASHES

Nuclear power plants were designed to withstand hurricanes, earthquakes, and other extreme events. But deliberate attacks by large airliners loaded with fuel, such as those that crashed into the World Trade Center and Pentagon, were not analyzed when design requirements for today's reactors were determined.[14] Concern about aircraft crashes was intensified by a taped interview shown September 10, 2002, on the Arab TV station al-Jazeera, which contained a statement that Al Qaeda initially planned to include a nuclear plant in its list of 2001 attack sites.

In light of the possibility that an air attack might penetrate the containment structure of a nuclear plant or a spent fuel storage facility, some interest groups have suggested that such an event could be followed by a meltdown or spent fuel fire and widespread radiation exposure. Nuclear industry spokespersons have countered by pointing out that relatively small, low-lying nuclear power plants are difficult targets for attack, and have argued that penetration of the containment is unlikely, and that even if such penetration occurred it probably would not reach the reactor vessel. Fires and explosions caused by an aircraft crash outside the reactor containment could disable systems required to cool the reactor core and spent fuel pools, and post-9/11 NRC regulations require nuclear plants to be able to mitigate such effects. According to former NRC Chairman Nils Diaz, NRC studies, which have not been released, "confirm that the likelihood of both damaging the reactor core and releasing radioactivity that could affect public health and safety is low."[15] However, groups critical of nuclear power consider NRC's mitigation requirements for fires and explosions to be insufficient to protect against aircraft crashes.[16]

NRC proposed in October 2007 to amend its regulations to require newly designed power reactors to take into account the potential effects of the impact of a large commercial aircraft. [17] As discussed in the previous section, NRC considers an aircraft attack to be beyond the design basis threat that plants must be able to withstand, so the requirements of the proposed rule were intended to provide an additional margin of safety. The proposed rule was drafted to affect only new reactor designs not previously certified by NRC, because the previous designs were still considered adequately safe. Nevertheless, Westinghouse submitted changes in the certified design of its AP1000 reactor to NRC on May 29, 2007, proposing to line the inside and outside of the reactor's concrete shield building with steel plates to increase resistance to aircraft penetration.[18]

Under NRC's 2007 proposed rule, applicants for new certified designs or for new reactor licenses using uncertified designs would have been required to assess the effects that a large aircraft crash would have on the proposed facilities. Each applicant would then describe how the plant's design features, capabilities, and operations would avoid or mitigate the effects of such a crash, particularly on core cooling, containment integrity, and spent fuel storage pools.

In response to comments, the NRC staff proposed in October 2008 that the aircraft impact assessments be conducted by all new reactors, including those using previously certified designs.[19] The NRC Commissioners, in a 3-1 vote, approved the change February 17, 2009, [20] and it was published in the Federal Register June 12, 2009.[21] The new rule added specific design requirements that all new reactors must meet:

> Each applicant subject to this section shall perform a design-specific assessment of the effects on the facility of the impact of a large, commercial aircraft. Using realistic analyses, the applicant shall identify and incorporate into the design those design features and functional capabilities to show that, with reduced use of operator actions:
>
> a) the reactor core remains cooled, or the containment remains intact; and
> b) spent fuel cooling or spent fuel pool integrity is maintained.

As noted above, NRC rejected proposals that existing reactors—in addition to new reactors—be required to protect against aircraft crashes, such as by adding "beamhenge" barriers. NRC determined that damage from aircraft crashes at existing reactors would be sufficiently mitigated by measures that had already been required at all reactors. In 2002, NRC ordered all nuclear power plants to develop strategies to mitigate the effects of large fires and explosions that could result from aircraft crashes or other causes.[22] As part of a broad security rulemaking effort, NRC proposed in October 2006 to incorporate the 2002 order on fire and explosion strategies into its security regulations (10 CFR Part 73).[23] In response to comments, NRC published a supplemental proposed rule in April 2008 to move the fire and explosion requirements into its reactor licensing regulations at 10 CFR Part 50, along with requirements that reactors establish procedures for responding to specific aircraft threat notifications.[24] Those regulations received final approval by the NRC Commissioners December 17, 2008,[25] and were published in the Federal Register March 27, 2009.[26] A key provision in the new rule states:

Each licensee shall develop and implement guidance and strategies intended to maintain or restore core cooling, containment, and spent fuel pool cooling capabilities under the circumstances associated with loss of large areas of the plant due to explosions or fire, to include strategies in the following areas:

(i) Fire fighting;
(ii) Operations to mitigate fuel damage; and
(iii) Actions to minimize radiological release.[27]

SPENT FUEL STORAGE

When no longer capable of efficiently sustaining a nuclear chain reaction, highly radioactive "spent" nuclear fuel is removed from the reactor and stored in a pool of water in the reactor building and at many sites later transferred to dry casks on the plant grounds. Because both types of storage are located outside the reactor containment structure, particular concern has been raised about the vulnerability of spent fuel to attack by aircraft or other means. If terrorists could breach a spent fuel pool's concrete walls and drain the cooling water, or disable cooling systems so that cooling water evaporated, the spent fuel's zirconium cladding could overheat and catch fire.

The National Academy of Sciences (NAS) released a report in April 2005 that found that "successful terrorist attacks on spent fuel pools, though difficult, are possible," and that "if an attack leads to a propagating zirconium cladding fire, it could result in the release of large amounts of radioactive material." NAS recommended that the hottest spent fuel be interspersed with cooler spent fuel to reduce the likelihood of fire, and that water-spray systems be installed to cool spent fuel if pool water were lost. The report also called for NRC to conduct more analysis of the issue and consider earlier movement of spent fuel from pools into dry storage.[28]

NRC agreed with some of findings of the NAS study but disagreed in several areas. In a report to Congress in response to the NAS report, NRC stated:

In summary, the NRC believes based on information developed in NRC vulnerability assessments, that the Committee has identified some scenarios that are unreasonable. The NRC also disagrees with some NAS recommendations and its conclusion lacks a sound technical basis. The

NAS finding that earlier movement of spent fuel from pools into dry storage would be prudent is one such example.[29]

NRC conducted the site-specific analyses recommended by NAS with funding provided by the FY2006 Energy and Water Development Appropriations Act (P.L. 109-103, H.Rept. 109-275). NRC's March 2009 regulations cited above include "spent fuel pool cooling capabilities" as a function that must be addressed by nuclear plants' mitigation strategies for large fires and explosions. Protection of spent fuel cooling also is included in the design requirements for new reactors under NRC's June 2009 aircraft impact regulations.

When the Fukushima accident raised additional concern about spent fuel pool vulnerability, NRC staff examined whether further studies should be conducted on expediting the transfer of spent fuel into dry storage. The staff reported to the Commission on November 12, 2013, that "the expedited transfer of spent fuel to dry cask storage would provide only a minor or limited safety benefit, and that its expected implementation costs would not be warranted."[30] The November 2013 study was based partly on an NRC analysis of spent fuel pool earthquake vulnerability released in October 2013. The October analysis said that spent fuel in a pool that was drained could be hot enough to catch fire during the first few months after being removed from the reactor. After that period, the spent fuel would have cooled too much to ignite, according to the study. Overall, the risk to the public from spent fuel pool damage was found to be extremely low. However, several comments to the October study, including from the New York attorney general's office, contended that NRC had underestimated the spent fuel pool accident risk.[31]

NRC has long contended that the potential effects of terrorist attacks are not "reasonably foreseeable" impacts that must be included in environmental studies for proposed spent fuel storage and other nuclear facilities. However, the U.S. Court of Appeals for the 9th Circuit ruled in June 2006 that terrorist attacks must be included in the environmental study of a dry storage facility at California's Diablo Canyon nuclear plant. NRC reissued the Diablo Canyon study May 29, 2007, to comply with the court ruling, but it did not include terrorism in other recent environmental studies outside the jurisdiction of the 9th Circuit.[32] The U.S. Court of Appeals for the 3rd Circuit subsequently ruled that NRC did not have to consider the impact of terrorist attacks in the license renewal application for the Oyster Creek plant in New Jersey.[33]

Long-term management of spent nuclear fuel is currently undergoing review, but spent fuel stored at reactor sites is expected to be moved eventually to central storage, permanent disposal, or reprocessing facilities. Large-scale transportation campaigns would increase public attention to NRC transportation security requirements and related security issues.

FORCE-ON-FORCE EXERCISES

EPACT05 codified an NRC requirement that each nuclear power plant conduct security exercises every three years to test its ability to defend against the design basis threat. In these "force-onforce" exercises, closely monitored and evaluated by NRC, a mock adversary force from outside the plant attempts to penetrate the plant's vital area and simulate damage to a "target set" of key safety components. Actual damage to such components could result in radioactive releases from the plant.

Participants in the tightly controlled exercises carry weapons modified to fire only blanks and laser bursts to simulate bullets, and they wear laser sensors to indicate hits. Other weapons and explosives, as well as destruction or breaching of physical security barriers, may also be simulated. While one squad of the plant's guard force is participating in a force-on-force exercise, another squad is also on duty to maintain normal plant security. Plant defenders know that a mock attack will take place sometime during a specific period of several hours, but they do not know what the attack scenario will be. Multiple attack scenarios are conducted over several days of exercises.

Full implementation of the force-on-force program began in late 2004. Standard procedures and other requirements have been developed for using the force-on-force exercises to evaluate plant security and as a basis for taking regulatory enforcement action. Many tradeoffs are necessary to make the exercises as realistic and consistent as possible without endangering participants or regular plant operations and security.

NRC required the nuclear industry to develop and train, under NRC standards, a "composite adversary force" made up of security officers from many plants to simulate terrorist attacks in all force-on-force exercises conducted after October 2004. However, in September 2004 testimony, the Government Accountability Office (GAO) criticized the industry's selection of Wackenhut (now G4S Regulated Security Solutions), a security company that guards many U.S. nuclear plants, to manage the adversary force, including non-Wackenhut employees. In addition to raising "questions about the force's independence," GAO noted that Wackenhut had been accused of cheating on previous force-on-force exercises by the Department of Energy.[34] Exelon terminated its security contracts with Wackenhut in late 2007 after guards at the Peach Bottom reactor in York County, PA, were discovered sleeping while on duty.[35]

EPACT05 requires NRC to "mitigate any potential conflict of interest that could influence the results of a force-on-force exercise, as the Commission determines to be necessary and appropriate." NRC prohibits officers in the adversary force from participating in exercises at their home plants. As in previous years, NRC's 2012 annual security report to Congress found that the industry adversary teams "continued to meet expectations for a credible, well-trained mock adversary force."[36]

During calendar year 2012, NRC conducted 23 force-on-force (FOF) inspections at 22 commercial nuclear plants and one fuel cycle facility. Eleven of those inspections found performance deficiencies: 19 with low significance (green findings), one "greater than green" finding, and three severity level IV (least serious) violations. One exercise resulted in the simulated destruction of or damage to a complete "target set" of vital plant components that were under mock attack.

According to NRC, "Any significant deficiencies in the protective strategy identified during FOF exercises are promptly reviewed and corrected." If a complete target set is simulated to be destroyed or damaged, "compensatory measures," such as extra security patrols, are required until permanent solutions can be implemented. Usually, "compensatory measures will be put in place before the NRC inspection team leaves the site area." Follow-up force-on-force exercises are sometimes conducted to verify that the necessary security improvements have been made.[37]

EMERGENCY PREPAREDNESS AND RESPONSE

After the 1979 accident at the Three Mile Island nuclear plant near Harrisburg, PA, Congress required that all nuclear power plants be covered by emergency plans. NRC requires each plant to have an Emergency Planning Zone (EPZ) with an approximately 10-mile radius. Within the 10- mile EPZ, the plant operator must maintain warning systems and evacuation plans. Every two years, each plant must conduct full-scale emergency response exercises that are evaluated by NRC and the Federal Emergency Management Agency (FEMA). Within 50 miles of each plant, plans must be in place to prevent the ingestion of radioactive material after an accident, such as through bans on contaminated food and water.[38]

In light of heightened concern about terrorist attacks since 9/11, proposals have been made to expand the EPZ to include larger population centers. NRC determined that the 10-mile EPZ remained adequate, but it issued a bulletin in July 2005 identifying enhancements for emergency response plans in the case of "security-based events at a nuclear power plant."[39] The Fukushima disaster raised further questions about the adequacy of the 10-mile EPZ, particularly because the Fukushima evacuation zone extended as far as 25 miles from the plant.[40] An NRC task force on Fukushima did not recommend expansion of the EPZ, concluding that emergency planning "enhancements" should be evaluated as additional information becomes available.[41]

The potential release of radioactive iodine during a nuclear incident is a particular concern, because iodine tends to concentrate in the thyroid gland of persons exposed to it. Emergency plans in many states include distribution of iodine pills to the population within the EPZ. Taking non-radioactive iodine before exposure would prevent absorption of radioactive iodine but would afford no protection against other radioactive elements. In 2002, NRC began providing iodine pills to states requesting them for populations within the 10-mile EPZ.

NRC completed one of its final regulatory responses to the 9/11 attacks by issuing wide-ranging revisions to its emergency preparedness regulations on November 23, 2011. The changes include requirements that plant staff with emergency duties not have potentially interfering responsibilities; that hostile actions be included as a type of emergency that requires plant coordination with local, state, and federal agencies; that emergency plans include protection of plant personnel from hostile action; and that hostile actions be included in emergency preparedness drills. Some non-security changes were also included, such as a requirement that evacuation time estimates be updated after every census or whenever major population changes occur.[42]

CYBERSECURITY

Existing U.S. nuclear power reactors, designed in the 1960s and '70s, are controlled primarily by analog systems that are resistant to cyberattack. However, new reactors are being designed with digital controls, and existing analog plants increasingly rely on digital computers to run auxiliary and monitoring systems. This increasing use of digital systems in nuclear power plants, along with post 9/11 security concerns and at least one "worm" infection at a U.S. reactor,[43] have prompted increased NRC attention to cybersecurity.

A year after the 9/11 attacks, NRC issued an order that included cyberattacks among the threats that nuclear plants would be required to defend against. Additional guidance for dealing with cyber threats was released during the next several years, and NRC issued formal cybersecurity regulations in March 2009 ("Protection of Digital Computer and Communications Systems and Networks," 10 CFR 73.54). NRC published a regulatory guide for the program in January 2010.[44]

NRC's cybersecurity regulations require each nuclear power plant to submit a cybersecurity plan and implementation schedule. The plan must provide "high assurance" that digital computer and communications systems that perform the following functions will provide adequate protection against design basis attacks:

- Functions that are safety-related or important to safety;
- Security functions;
- Emergency preparedness functions, including offsite communications; and
- Support systems and equipment that, if compromised, would adversely affect safety, security, or emergency preparedness functions.

NRC began inspecting the implementation of nuclear plant cybersecurity plans in January 2013. The inspections are part of NRC's Cyber Security Oversight Program, which is being incorporated into the existing Reactor Oversight Program.[45]

Nuclear power plants are also required by the Federal Energy Regulatory Commission (FERC) to comply with cybersecurity standards issued by the North American Electric Reliability Corporation (NERC). However, nuclear plant computer systems that are covered by NRC security regulations are exempt from the NERC standards. As a result, the NERC standards apply mostly to "balance of plant" (non-reactor) systems at nuclear power plants.[46]

PLANT SECURITY PERSONNEL

After video recordings of inattentive security officers at the Peach Bottom (PA) nuclear power plant were aired on local television, an NRC inspection in late September 2007 confirmed that there had been multiple occasions on which multiple security officers were inattentive.[47] However, after a follow-up inspection into security issues at the Peach Bottom plant, run by Exelon Nuclear, the NRC concluded that the plant's security program had not been significantly degraded as a result of the guards' inattentiveness. NRC issued a bulletin December 12, 2007, requiring all nuclear power plants to provide written descriptions of their "managerial controls to deter and address inattentiveness and complicity among licensee security personnel."[48]

The incident drew harsh criticism from the House Committee on Energy and Commerce. "The NRC's stunning failure to act on credible allegations of sleeping security guards, coupled with its unwillingness to protect the whistleblower who uncovered the problem, raises troubling questions," said Representative John D. Dingell, then-Chairman of the Committee.[49] NRC proposed a $65,000 fine on Exelon Nuclear on January 6, 2009.[50]

NRC modified its standards for plant security personnel as part of a major revision of its security regulations published March 27, 2009 (described in the next section). The standards cover education and training requirements, criminal background checks, and mental and physical qualifications.[51]

OVERVIEW OF NRC ACTIONS AFTER 9/11

Following the 9/11 terrorist attacks, NRC conducted a "top-to-bottom" review of its nuclear power plant security requirements. On February 25, 2002, the agency issued "interim compensatory security measures" to deal with the "generalized high-level threat environment" that continued to exist, and on January 7, 2003, it issued regulatory orders that tightened nuclear plant access. On April 29, 2003, NRC issued orders to restrict security officer work hours, establish new security force training and qualification requirements, and increase the DBT that nuclear security forces must be able to defend against, as discussed previously.

In October 2006, NRC proposed to amend the security regulations and add new security requirements that would codify the series of orders issued after 9/11 and respond to requirements in the Energy Policy Act of 2005.[52] The new security regulations were approved by the NRC Commissioners on December 17, 2008, and published March 27, 2009:[53]

- *Safety and Security Interface.* Explicit requirements were established for nuclear plants to ensure that necessary security measures do not compromise plant safety.
- *Mixed-Oxide Fuel.* Enhanced physical security requirements were established to prevent theft or diversion of plutonium-bearing mixed-oxide (MOX) fuel.
- *Cybersecurity.* Nuclear plants must submit security plans that describe how digital computer and communications systems and safety-related networks are protected from cyberattacks, as discussed in the cybersecurity section above.
- *Aircraft Attack Mitigative Strategies and Response.* As discussed in the earlier section on vulnerability to aircraft crashes, nuclear plants must prepare strategies for responding to warnings of an aircraft attack and for mitigating the effects of large explosions and fires.
- *Plant Access Authorization.* Nuclear plants must implement more rigorous programs for authorizing access, including enhanced psychological assessments and behavioral observation.
- *Security Personnel Training and Qualification.* Modifications to security personnel requirements include additional physical fitness standards, increased minimum qualification scores for mandatory personnel tests, and requirements for on-the-job training.
- *Physical Security Enhancements.* New requirements are intended to ensure the availability of backup security command centers, uninterruptible power supplies to detection systems, enhanced video capability, and protection from waterborne vehicles.

A proposal by NRC staff to release more details about the results of nuclear plant security inspections was defeated by the NRC Commissioners in a 2-2 vote on January 21, 2009. Under current policy, NRC announces after a security inspection whether any violations that were found were of low safety significance or moderate-or-higher safety significance. Critics of the current policy contend that the public needs more detail to be assured of plant security. The policy's supporters counter that greater information about security inspection findings could inadvertently provide useful information to terrorists.[54]

SELECTED LEGISLATION IN THE 113TH CONGRESS

H.R. 1700 (Engel)

Nuclear Disaster Preparedness Act. Requires the President, in consultation with the Secretary of Energy, the Secretary of Defense, the Administrator of the Environmental Protection Agency, the Nuclear Regulatory Commission, the Secretary of Homeland Security, and the Administrator of the Federal Emergency Management Agency, to issue guidance on federal response to nuclear disasters resulting from terrorism and other causes. Introduced April 24, 2013; referred to Committee on Transportation and Infrastructure.

H.R. 3354 (Engel)

Dry Cask Storage Act. Requires spent fuel at nuclear power plants to be moved from spent fuel pools to dry casks after it has sufficiently cooled. Costs of the fuel transfers would be offset by a reduction in nuclear waste fees owed to the federal government. Introduced October 28, 2013; referred to Committee on Energy and Commerce.

End Notes

[1] Homeland Security Presidential Directive 7: Critical Infrastructure Identification, Prioritization, and Protection, December 17, 2003, http://www.dhs.gov/xabout/laws/gc_1214597989952.shtm#1.

[2] General NRC requirements for nuclear power plant security can be found in 10 C.F.R. 73.55.

[3] 10 C.F.R. §73.1.

[4] It was feared that Cuba might launch an attack on Florida reactors. Government Accountability Office, Nuclear Power Plants—Efforts Made to Upgrade Security, but the Nuclear Regulatory Commission's Design Basis Threat Process Should Be Improved (GAO-06-388), March 2006, p. 2. Regulations at 10 CFR 50.13.

[5] 10 C.F.R. §50.13. Attacks and destructive acts by enemies of the United States; and defense activities.

[6] Nuclear Regulatory Commission, "Design Basis Threat," 72 Federal Register 12714, March 19, 2007.

[7] Federal Register, May 7, 2003 (vol. 68, no. 88). NRC, All Operating Power Reactor Licensees; Order Modifying Licenses.

[8] P.L. 109-58, Title VI, Subtitle D—Nuclear Security (Sections 651-657). Section 651 adds Atomic Energy Act Section 170E. Design Basis Threat Rulemaking.

[9] Federal Register, March 19, 2007 (vol. 72, no. 52), NRC, Design Basis Threat, Final Rule, pp. 12705-12727.

[10] NRC, "NRC Approves Final Rule Amending Security Requirements," News Release No. 07-012, January 29, 2007.

[11] Edwin S. Lyman, "Security Since September 11th," Nuclear Engineering International, March 2010, pp. 14-19.

[12] 10 C.F.R. 73.55 (k)(5)(ii) states: "The number of armed responders shall not be less than ten (10)." The previous requirement, in 10 C.F.R. 73.55 (h)(3), stated: "The total number of guards, and armed, trained personnel immediately available at the facility to fulfill these response requirements shall nominally be ten (10), unless specifically required otherwise on a case by case basis by the Commission; however, this number may not be reduced to less than five (5) guards." The change was made in NRC final regulations published in March 2009, op. cit.

[13] Doug Walters, "Security Since March," Nuclear Engineering International, May 2010.

[14] Meserve, Richard A., NRC Chairman, "Research: Strengthening the Foundation of the Nuclear Industry," Speech to Nuclear Safety Research Conference, October 29, 2002.

[15] Letter from NRC Chairman Nils J. Diaz to Secretary of Homeland Security Tom Ridge, September 8, 2004.

[16] Committee to Bridge the Gap, Nuclear Information and Resource Service, and Public Citizen, "NRC Votes Against Requiring Reactors to be Protected from Air Attacks or Large Numbers of Attackers," press release, January 29, 2007, http://www.committeetobridge thegap.org/pressrelease/012907release.html.

[17] Federal Register, October 3, 2007 (vol. 72, no. 191), Consideration of Aircraft Impacts for New Nuclear Power Reactor Designs.

[18] MacLachlan, Ann, "Westinghouse Changes AP1000 Design to Improve Plane Crash Resistance," Nucleonics Week, June 21, 2007.

[19] Nuclear Regulatory Commission, Final Rule—Consideration of Aircraft Impacts for New Nuclear Power Reactors, Rulemaking Issue Affirmation, SECY-08-0152, October 15, 2008.

[20] Nuclear Regulatory Commission, Final Rule—Consideration of Aircraft Impacts for New Nuclear Power Reactors, Commission Voting Record, SECY-08-0152, February 17, 2009.

[21] Nuclear Regulatory Commission, Consideration of Aircraft Impacts for New Nuclear Power Reactors, Final Rule, 74 Federal Register 28111, June 12, 2009. This provision is codified at 10 CFR 50.150.

[22] Nuclear Regulatory Commission, Final Rule—Consideration of Aircraft Impacts for New Nuclear Power Reactors, Rulemaking Issue Affirmation, SECY-08-0152, October 15, 2008, p. 2.

[23] Nuclear Regulatory Commission, "Power Reactor Security Requirements, Proposed Rule," 71 Federal Register 62664, October 26, 2006.

[24] Nuclear Regulatory Commission, "Power Reactor Security Requirements, Supplemental Proposed Rule," 73 Federal Register 19443, April 10, 2008.

[25] Nuclear Regulatory Commission, "NRC Approves Final Rule Expanding Security Requirements for Nuclear Power Plants," press release, December 17, 2008, http://www.nrc.gov/reading-rm/doc-collections/news/2008/08-227.html.

[26] Nuclear Regulatory Commission, Power Reactor Security Requirements, Final Rule, 74 Federal Register 13925, March 27, 2009.

[27] 10 CFR 50.54(hh)(2).

[28] National Academy of Sciences, Board on Radioactive Waste Management, Safety and Security of Commercial Spent Nuclear Fuel Storage, Public Report (online version), released April 6, 2005.

[29] U.S. Nuclear Regulatory Commission Report to Congress on the National Academy of Sciences Study on the Safety and Security of Commercial Spent Nuclear Fuel Storage, March 2005, p. iii, http://www.nrc.gov/reading-rm/doccollections/congress-docs/correspondence/2005/domenici-03142005.pdf.

[30] Satorius, Mark A., NRC executive director for operations, memo to NRC commissioners, COMSECY-13-0030, November 12, 2013, http://www.nrc.gov/reading-rm/doc-collections/ commission/comm-secy/2013/2013- 0030comscy.pdf.

[31] NRC Office of Nuclear Regulatory Research, Consequence Study of a Beyond-Design-Basis Earthquake Affecting the Spent Fuel Pool for a U.S. Mark I Boiling Water Reactor, October 2013, http://pbadupws.nrc.gov/docs/ML1325/ ML13256A342.pdf.

[32] Beattie, Jeff, "NRC Takes Two Roads on Terror Review Issue," Energy Daily, February 27, 2007.

[33] U.S. Court of Appeals for the Third Circuit, New Jersey Department of Environmental Protection v. U.S. Nuclear Regulatory Commission, March 31, 2009, http://www.ca3.uscourts.gov/opinarch/072271p.pdf.

[34] GAO. "Nuclear Regulatory Commission: Preliminary Observations on Efforts to Improve Security at Nuclear Power Plants." Statement of Jim Wells, Director, Natural Resources and Environment, to the Subcommittee on National Security, Emerging Threats, and International Relations, House Committee on Government Reform. September 14, 2004. p. 14.

[35] Washington Post, "Executive Resigns in Storm Over Sleeping Guards," January 10, 2008.

[36] Nuclear Regulatory Commission, Office of Nuclear Security and Incident Response, Report to Congress on the Security Inspection Program for Commercial Power Reactor and Category 1 Fuel Cycle Facilities: Results and Status Update; Annual Report for Calendar Year 2012, NUREG-1885, Rev. 6, July 2013, p. 11, http://pbadupws.nrc.gov/docs/ ML1321/ ML13210A062.pdf

[37] NRC, Report to Congress on the Security Inspection Program, op. cit., p. 9.

[38] NRC, "Emergency Preparedness & Response," web page, December 19, 2013, http://www .nrc.gov/about-nrc/emergpreparedness.html.

[39] NRC Office of Nuclear Security and Incident Response, Emergency Preparedness and Response Actions for Security-Based Events, NRC Bulletin 2005-02, July 18, 2005, http://www.nrc.gov/reading-rm/doc-collections/gencomm/bulletins/2005/bl200502.pdf.

[40] Japanese Ministry of Economy, Trade, and Industry, "Restricted Areas and Areas to Which Evacuation Orders Have Been Issued," May 7, 2013, http://www.meti.go.jp/ english/ earthquake/nuclear/roadmap/pdf/130507_assistance.pdf.

[41] NRC Near-Term Task Force Review of Insights from the Fukushima Dai-ichi Accident, Recommendations for Enhancing Reactor Safety in the 21st Century, July 12, 2011, pp. 60-61, http://pbadupws.nrc.gov/docs/ML1118/ ML111861807.pdf.

[42] NRC, "Enhancements to Emergency Preparedness Regulations," final rule, Federal Register, November 23, 2011, p. 72560.

[43] Kesler, Brent, "The Vulnerability of Nuclear Facilities to Cyber Attack," Strategic Insights, spring 2001, p. 15, http://www.nps.edu/Academics/Centers/CCC/Research-Publications /StrategicInsights/2011/Apr/SI-v10-i1_Kesler.pdf.

[44] NRC, "Cyber Security Programs for Nuclear Facilities," Regulatory Guide 5.71, January 2010, http://pbadupws.nrc.gov/docs/ML0903/ML090340159.pdf.

[45] NRC, Report to Congress on the Security Inspection Program, op. cit., p. 7.

[46] Federal Energy Regulatory Commission, "Mandatory Reliability Standards for Critical Infrastructure Protection: Order Addressing Compliance Filing and Approving Implementation Plan," Docket No. RM06-22-011, March 18, 2010, http://www.nerc.com/ files/CIP_ImplementationOrder-03182010.pdf.

[47] NRC, NRC Commences Follow-up Security Inspection at Peach Bottom, November 5, 2007, http://www.nrc.gov/ reading-rm/doc-collections/news/2007/07-057.i.html.

[48] NRC, Security Officer Attentiveness, NRC Bulletin 2007-1, Washington, DC, December 12, 2007.

[49] Committee on Energy and Commerce, Energy and Commerce Committee to Probe Breakdowns in NRC Oversight, January 7, 2008 http://energycommerce. house.gov/ Press_110/110nr149.shtml.

[50] NRC, "NRC Proposes $65,000 Fine for Violations Associated with Inattentive Security Guards at Peach Bottom Nuclear Plant," press release, January 6, 2009, http://www.nrc.gov/ reading-rm/doc-collections/news/2009/09- 001.i.html.

[51] 10 Code of Federal Regulations Part 73, Appendix B—General Criteria for Security Personnel, http://www.nrc.gov/ reading-rm/doc-collections/cfr/part073/part073-appb.html.

[52] Federal Register, October 26, 2006 (vol. 71, no. 207), NRC, Power Reactor Security Requirements, Proposed Rule.

[53] Federal Register, March 27, 2009, op. cit.

[54] Jenny Weil, "Commissioners Reach Stalemate on Security-Related Amendment," Inside NRC, February 2, 2009.

In: Challenges Facing the Nuclear Power Industry ISBN: 978-1-63483-618-0
Editor: Renee Olson

Chapter 5

EPA'S PROPOSED GREENHOUSE GAS REGULATIONS: IMPLICATIONS FOR THE ELECTRIC POWER SECTOR*

Richard J. Campbell

SUMMARY

The Environmental Protection Agency (EPA) has proposed regulations to reduce greenhouse gas (GHG) emissions from existing power plants. EPA believes that its proposed Clean Power Plan (CPP) will "protect public health, move the United States towards a cleaner environment, and fight climate change while supplying Americans with reliable and affordable power." Burning fossil fuels to produce electricity results in the release of carbon dioxide, and represents the largest source of GHG emissions in the United States. Under its proposed plan, EPA believes it will be possible to lower the carbon intensity of power generation in the United States by approximately 30% in 2030 from carbon dioxide emissions levels in 2005. To achieve this goal, EPA is giving each state a numerical carbon reduction target, based on the state's existing power generation portfolio.

Under the Clean Air Act (CAA) section 111(d), the EPA must identify the best system of emission reductions (BSER) that is adequately

* This is an edited, reformatted and augmented version of a Congressional Research Service publication R43621, prepared for Members and Committees of Congress, dated June 23, 2014.

demonstrated and available to reduce pollution. The regulations allow EPA to set goals, and give states the responsibility for creating compliance plans which meet those goals. EPA set the state-specific goals based on (1) measures which improve the efficiency of fossil-fueled power plants, (2) use of lower-emitting generation sources such as natural gas or fuel-switching to natural gas, and use of nuclear power, (3) demand-side efficiency, and (4) renewable electric generation. EPA has suggested these measures as four BSER "building blocks" or options for states to consider when choosing how to meet their specific GHG reduction goals. Under the proposed plan, however, states would have the flexibility to choose the most cost-effective strategies to meet the targets.

EPA's proposal for GHG reduction answers some questions from the electric power sector with regard to the timeframe, timeline, and choices that would be made available for compliance. The CPP proposal sets out a vision for a greater proportion of electricity coming from natural gas and renewable electric generation, and less from coal-fired power plants. However, some issues still remain unresolved with the potential implementation of the CPP.

Some observers say that EPA's CPP essentially proposes an environmental dispatch regime for power plant operation which could potentially result in increased electricity prices to consumers, depending on the generation resource mix employed. But EPA believes adoption of greater energy efficiency measures will actually reduce average retail electricity bills. Compliance may require more natural gas consumption to firm up variable renewable electric generation. Increasing the use of natural gas for power generation has resulted in some concerns, as deliverability and price volatility issues have emerged as recently as this past winter. Power companies, gas suppliers, and regulatory regimes are working on resolving these issues.

The electric utility industry values diversity in fuel choice options since reliance on one fuel or technology can leave electricity producers vulnerable to price and supply volatility. Also, state-specific compliance plans geared to individual state needs may complicate the coordination necessary for reliability purposes. But EPA expects coal to remain a substantial part of the U.S. energy mix through 2030, allowing goals of fuel diversity and generating resource mix to be maintained.

Many fossil-fueled power plants do more than just generate electricity. Some of the power plants scheduled for retirement provide ancillary services to the grid such as voltage support and frequency regulation. Additional retirements of coal-fired capacity resulting from implementing the proposal could impact reserve margins and even grid reliability during weather-related outages or periods of temperature extremes. Incidents of more extreme weather appear to be occurring, and will need to be planned for when considering the types of future generation which may be needed to assure electric system reliability.

Implementing compliance plans will not come without real costs or hard choices for the states and electric utilities that will have to work together. Potential implications for reliability and the ultimate financial costs of the CPP are not known but will become clearer as state compliance plans are filed, and implementation plans become known.

INTRODUCTION

The Environmental Protection Agency (EPA) has proposed regulations to reduce greenhouse gas[1] (GHG) emissions from existing power plants (also referred to as electric generating units or EGUs by EPA). EPA believes that its proposed Clean Power Plan (CPP)[2] will "protect public health, move the United States towards a cleaner environment, and fight climate change while supplying Americans with reliable and affordable power."[3] Carbon emissions are linked by many to anthropogenic climate change,[4] and the EPA cites the Obama Administration's intent to address climate change[5] concerns in its proposed Clean Power Plan to reduce carbon emissions.

Since carbon dioxide (CO_2) from fossil fuel combustion is the primary GHG, and fossil fuels are used for the majority of electric power generation, the focus of the proposed policies is on reducing carbon emissions from power plants. Under its proposed plan, EPA believes it will be possible to lower the CO_2 emissions from power generation in the United States by approximately 30% by 2030 compared to levels in 2005. To achieve this goal, EPA is giving each state a numerical carbon reduction target, based on the state's existing power generation portfolio, and EPA's estimate of the state's potential to reduce power demand.

EPA's guidelines allow a 120-day comment period on the proposal, and require states to file a compliance plan by June 2016. States that choose to join a regional carbon reduction plan would have until June 2018 to submit the plan. The compliance mechanism would require states to reduce the carbon emissions rate of power plants (i.e., overall tons of carbon dioxide emitted per each MegaWatt-hour (MWh) of electricity produced). Whether working individually or regionally, states would be allowed to choose an appropriate mix of generation using diverse fuels (including renewable electricity[6] and nuclear power), energy efficiency, and demand-side management to meet the goals and their own needs.[7]

This report presents an analysis of EPA's proposed plan. Electric utilities and other stakeholders will be analyzing the CPP to understand its provisions, and have until the close of the comment period[8] to provide input. It is possible that the CPP will be modified in response to relevant comments received. The implications of implementing a final CPP are thus unlikely to be known until after the states file their compliance plans which are due by June 2016.

The issues for Congress will be focused on the implications of the CPP on electric power system reliability, the costs of electric power to customers, and the future structure of the electric utility industry which could result from implementation of compliance plans.

BACKGROUND

Burning fossil fuels to produce electricity results in the release of carbon dioxide, and represents the largest source of GHG emissions in the United States. As shown in *Figure 1*, fossil fuel combustion was responsible for approximately 68% of electric power generation as of 2012.Coal was the fuel most used. Coal is also the fossil fuel which emits the most carbon dioxide per unit of electric power produced, averaging 216 pounds of carbon dioxide per million British thermal units (mmBTUs) of energy produced. By comparison, natural gas combustion releases about half the carbon emissions at 117 pounds of carbon dioxide per mmBTU of energy produced.[9]

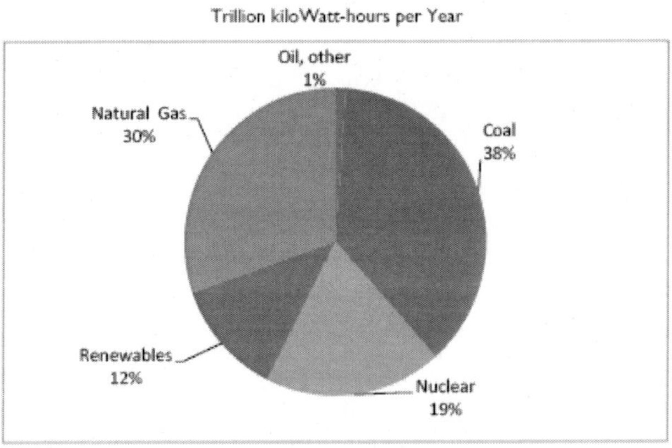

Source: DOE, Annual Energy Outlook, 2014 Early Release, December 16, 2013, http://www.eia.gov/forecasts/ aeo/er/early_elecgen.cfm.

Notes: Renewable electricity includes hydropower, wind, solar, and biomass power generation. "Other" includes other liquid fuels.

Figure 1. U.S. Electricity Generation by Fuel, 2012.

In a 2007 decision, the Supreme Court found in *Massachusetts vs. EPA*[10] that GHG emissions were air pollutants which could be regulated under the Clean Air Act (CAA).[11] EPA then moved in 2009 to declare that GHGs were a threat to public health in its "endangerment" finding, which served as a basis for its subsequent actions.[12] With regard to stationary sources of GHGs, EPA proposed standards in September 2013 for the control of carbon dioxide emissions from new electric power plants burning fossil fuels, under CAA section 111(b) regulations.

EPA has now released a proposal for reducing carbon dioxide emissions from existing power plants which burn fossil fuels. Under CAA 111(d), the EPA must establish a procedure under which states will submit plans establishing standards of performance for existing fossil fuel-fired power plants. The standards of performance are to reflect the degree of emissions limitation achievable through the application of the best system of emission reductions (BSER) that is adequately demonstrated and available to reduce pollution. The regulation under CAA 111(d) allows EPA to set goals, and gives states the responsibility for creating compliance plans which meet EPA's guidelines.

EPA'S PROPOSED PLAN FOR EXISTING COAL PLANTS[13]

Under the provisions of EPA's proposed Clean Power Plan, all existing fossil fuel-fired electric power generation plants must comply with new state-specific targets to reduce carbon emissions.[14] The combined state targets are expected to result in reducing carbon emissions from U.S. power generation approximately 30% by 2030 compared to carbon emissions levels in 2005.

EPA has designated four "building blocks" that it used to develop the state-specific GHG reduction goals:

1. *Improve the heat rate*[15] *of fossil-fueled power plants.* EPA suggests increasing power plant efficiency by equipment upgrades and improvements. Using less fossil fuel to create the same amount of electricity reduces carbon emissions. An average heat rate improvement of 6% is targeted for coal-fired power plants.[16]
2. *Increase use of low-emitting power sources.* EPA suggests more frequent use of power plants with lower carbon emissions resulting in less carbon pollution. Dispatching (i.e., scheduling the operation) of

higher efficiency natural gas combined cycle units more often is suggested.

3. *Use of more zero -emitting power sources.* EPA suggests expanding renewable electricity generation (such as zero-carbon emission wind and solar facilities), and nuclear power plants as a way to lower carbon emissions.
4. *Using electricity more efficiently.* EPA suggests energy efficiency as a way to reduce power demand, with a targeted increase in demand-side energy efficiency of 1.5% annually.

EPA recognizes that such options are already being used by some states to promote clean energy and efficiency goals. EPA considers a BSER to include measures which improve the efficiency of fossil-fueled power plants, use lower emitting generation sources such as natural gas or fuel-switching to natural gas, and use nuclear power and renewable electric generation. Under the proposed plan, EPA believes the states have the flexibility to choose the most cost-effective strategies to meet the targets.

States must file a plan by 2016 to comply with EPA's state-specific goals, unless they choose to join a multi-state regional plan. Each state has been given a unique carbon emissions reduction goal,[17] which EPA maintains offers broad flexibility in which to plan and achieve reductions in carbon emissions. *Option 1* would involve a higher level of deployment of the four building blocks over a longer timeframe (i.e., over 15 years to 2030). EPA also suggested an alternative path on which it is asking for comment in *Option 2*, which would allow a lower level of deployment of the four building blocks but over a shorter timeframe (i.e., over ten years to 2025). EPA has set interim and final goals for the state-specific goals under both compliance scenarios. They can meet the goals as individual states, or join together for regional solutions.

The state goals are given as an overall rate of carbon dioxide emissions intensity, that is, the amount of carbon dioxide emissions in pounds (lbs.) from fossil-fueled power plants *divided* by the amount of electricity generated in the state (from fossil-fired generation and low- or zero-carbon emitting power sources) in MWh. EPA estimates that the state-specific carbon reduction goals will result in the elimination of approximately 730 million metric tonnes[18] of carbon by 2030, resulting in a reduction of approximately 30% of carbon levels compared to 2005.[19]

DISCUSSION OF EPA'S PROPOSAL

Electric power generation in the United States differs regionally, and largely reflects local resources, fuel costs, and availability of fuel supplies.[20] EPA recognizes that it will take time to implement compliance solutions to meet its proposed carbon pollution reduction plan. EPA has attempted to provide flexibility for state compliance with its plan for reducing carbon emissions from existing fossil-fueled power plants.

> The EPA is also proposing to give states considerable flexibility with respect to the timeframes for plan development and implementation, with up to two or three years permitted for final plans to be submitted after the proposed GHG emission guidelines are finalized, and up to fifteen years for all emission reduction measures to be fully implemented.[21]

While 2005 has been mentioned in broader U.S. policy terms for reductions in GHG emissions to 2030, it is not the year that EPA has used in its emissions reduction calculation. EPA chose 2012 as the year from which to establish a baseline for emissions reduction since that was the year for which it has the most complete state emissions, net generation, and capacity data for all affected EGUs.

> For purposes of establishing state goals, historical (2012) electric generation data are used to apply each building block and develop each state's goal (expressed as an adjusted CO2 emission rate in lbs per MWh).[22]

Many regard this as beneficial for many states since U.S. GHG emissions overall have dropped 15% between 2005 and 2012.[23]

Existing State Clean Energy Programs

EPA also considers expanding state renewable electric programs and portfolio standards as tools to reduce GHG emissions. However, for purposes of establishing a baseline, existing hydropower is not included.

> Hydropower generation is excluded from this existing 2012 generation for purposes of quantifying BSER-related [renewable

> electricity] generation potential because building the methodology from a baseline that includes large amounts of existing hydropower generation could distort regional targets that are later applied to states lacking that existing hydropower capacity. The exclusion of pre-existing hydropower generation from the baseline of this target-setting framework does not prevent states from considering incremental hydropower generation from existing facilities (or later-built facilities) as an option for compliance with state goals.[24]

However, states will not receive credit for "early action" taken to reduce GHGs in the period prior to the timeframe of the proposal. Therefore, GHG reductions resulting from state renewable portfolio standards (and other similar measures) in the period from 2005 to 2012 will not count towards GHG reductions in the 2020 to 2030 timeframe. But since they are programs already in place, EPA considers these programs as helpful to meeting state-specific goals in the compliance timeframe.

> The EPA is also proposing that measures a state takes after the date of this proposal, or programs already in place, which result in CO2 emissions reduction during the 2020-2030 period, would apply toward achievement of the state's 2030 CO2 emissions goal. Thus, states with currently existing programs and policies, and states that put in new programs and policies early, will be better positioned to achieve the goals.[25]

While states must begin compliance with the proposed plan by 2020, full compliance is not required until 2030. EPA points out that its four building blocks are suggested as a framework for implementation; the states themselves will decide how to meet their goals. EPA also suggests that regional compliance strategies would be acceptable under its proposal.

> Each state will have the flexibility to design a program to meet its goal in a manner that reflects its particular circumstances and energy and environmental policy objectives. Each state can do so alone or can collaborate with other states on multi-state plans that may provide additional opportunities for cost savings and flexibility.[26]

In establishing the state-specific goals for GHG reduction, EPA believes it has taken into consideration the regional differences that exist in power generation types and resources. Each state's goal reflects the fact that a state's

CO_2 emissions are a result of how efficiently its fossil-fueled power plants operate, and how much they operate. Under the state-specific goals, the highest allowed rate of GHG emissions is 1,783 lbs. of CO_2 per MWh of electricity produced for North Dakota, while the lowest emissions rate is 215 lbs. of CO_2 per MWh in Washington state,[27] with the apparent disparity in state goals reflecting the existing underlying generation mix in the states.

Best System of Emissions Reduction

EPA has modeled the opportunities for heat rate improvement, dispatch of more gas and less coal, increased renewable and nuclear generation, and end-use energy efficiency. The agency considers these actions representing the "Best System of Emissions Reductions" which is adequately demonstrated.[28]

> Overall, the BSER proposed here is based on a range of measures that fall into four main categories, or "building blocks," which comprise improved operations at EGUs, dispatching lower-emitting EGUs, and zero-emitting energy sources, and end-use energy efficiency. All of these measures have been amply demonstrated via their current widespread use by utilities and states.
>
> The proposed guidelines are structured so that states would not be required to use each and every one of the measures that the EPA determines constitute the BSER or to apply any one of those measures to the same extent that the EPA determines is achievable at reasonable cost. Instead, in developing its plan, each state will have the flexibility to select the measure or combination of measures it prefers in order to achieve its CO_2 emission reduction goal.[29]
>
> As part of the BSER determination, the EPA considered the impacts that implementation of the emission reductions [based on the combination of the blocks] would have on the cost of electricity and electricity system reliability.... Importantly, the proposed BSER, expressed as a numeric goal for each state, provides states with the flexibility to determine how to achieve the reductions (i.e., greater reductions from one building block and less from another) and to adjust the timing in which reductions are achieved, in order to address key issues such as cost to consumers, electricity system reliability and the remaining useful life of existing generation assets.[30]
>
> The EPA is proposing to evaluate and approve state plans based on four general criteria: 1) enforceable measures that reduce EGU CO2

> emissions; 2) projected achievement of emission performance equivalent to the goals established by the EPA, on a timeline equivalent to that in the emission guidelines; 3) quantifiable and verifiable emission reductions; and 4) a process for biennial reporting on plan implementation, progress toward achieving CO2 goals, and implementation of corrective actions, if necessary.[31]

In devising a compliance strategy, EPA proposes to allow each state to design a GHG reduction program using strategies or technologies the state selects.

> To meet its goal, each state will be able to design programs that use the measures it selects, and these may include the combination of building blocks most relevant to its specific circumstances and policy preferences. States may also identify technologies or strategies that are not explicitly mentioned in any of the four building blocks and may use those technologies or strategies as part of their overall plans (e.g., market-based trading programs or construction of new natural combined cycle units or nuclear plants).[32]

EPA believes that shifting electric power production from coal-fired power plants to natural gas combined-cycle generation (NGCC) represents a major opportunity to reduce GHG emissions due to the greater overall efficiency of the newer NGCC generation fleet especially as compared to older coal-fired power plants, and the current underutilization of NGCC generation capacity.

> Our analysis indicated that the potential CO2 reductions available through re-dispatch from steam EGUs to NGCC units are substantial. As of 2012, there was approximately 245 GW of NGCC capacity in the United States, 196 GW of which was placed in service between 2000 and 2012. In 2012, the average utilization rate of U.S. NGCC capacity was 46 percent, well below the utilization rates the units are capable of achieving. In 2012, approximately 10 percent of NGCC plants operated at annual utilization rates of 70 percent or higher, and 19 percent of NGCC units operated at utilization rates of at least 70 percent over the summer season. Average reported availability generally exceeds 85 percent. We recognize that the ability to increase NGCC utilization rates may also be affected by infrastructure and system considerations, such as limits on the ability of the natural gas industry to produce and deliver the increased

quantities of natural gas, the ability of steam EGUs to reduce generation while remaining ready to supply electricity when needed in peak demand hours, and the ability of the electric transmission system to accommodate the changed geographic pattern of generation.[33]

The amount of re-dispatch from coal-fired EGUs to NGCC units that takes place as a result of this competition is highly relevant to overall power sector GHG emissions, because a typical NGCC unit produces less than half as much CO2 per MWh of electricity generated as a typical coal-fired EGU.[34]

Under its proposal, EPA has suggested a range of equipment upgrades and improvement options to increase heat rates (specifically at existing coal-fired power plants), which it believes represents cost-effective opportunities to reduce GHG emissions.

Our assessment of heat rate improvements showed that these measures would achieve CO2 emission reductions at low costs, although compared to other measures, the available reductions were relatively limited in quantity. Specifically, our analysis indicated that average CO2 emission reductions of 1.3 to 6.7 percent could be achieved by coal-fired steam EGUs through adoption of best practices, and that additional average reductions of up to four percent could be achieved through equipment upgrades. Heat rate improvements pay for themselves at least in part through reductions in fuel costs, generally making this a relatively inexpensive approach for reducing CO2 emissions. We estimated that CO2 reductions of between four and six percent from overall heat rate improvements could be achieved on average across the nation's fleet of coal-fired steam EGUs for net costs in a range of $6 to $12 per metric ton.[35]

Although heat rate improvements have the potential to reduce CO2 emissions from all types of affected EGUs, the EPA's analysis indicates the potential is significantly greater for coal-fired steam EGUs than for other EGUs, and for purposes of determining the best system of emission reduction at this time EPA is conservatively proposing to base its estimate of CO2 emission reductions from heat rate improvements on coal-fired steam EGUs only.[36]

Fuel Switching

EPA sees the opportunity for coal to natural gas conversions at existing power plants (rather than co-firing coal and natural gas together) as having a great potential to reduce GHGs. While converting a coal-fired power plant to natural gas would be considered a higher cost option compared to other heat rate improvements, EPA believes the incremental difference in the cost of fuel would be the most significant cost component.

> Natural gas co-firing or conversion at coal-fired steam EGUs offers greater potential CO2 emission reductions than heat rate improvements, but at a higher cost (although less than the cost of applying CCS [Carbon Capture and Storage] technology). Because natural gas contains less carbon than an energy-equivalent quantity of coal, converting a coal-fired steam EGU to burn only natural gas would reduce the unit's CO2 emissions by approximately 40 percent. The CO2 reductions are generally proportional to the amount of gas substituted for coal, so if an EGU continued to burn mostly coal while co-firing natural gas as, for example, 10 percent of the EGU's total heat input, the CO2 emission reductions would be approximately four percent. The EPA determined that the most significant cost associated with natural gas conversion or co-firing is likely to be the incremental cost of natural gas relative to the cost of coal. [37]

Nuclear Power

Another higher cost option would be the construction of new nuclear power plants. However, EPA views the completion of nuclear units currently under construction, and avoiding the "premature" retirement of an estimated 6% of existing nuclear capacity, to be important to GHG reduction goals in some states.[38]

> Policies that encourage development of renewable energy capacity and discourage premature retirement of nuclear capacity could be useful elements of CO_2 reduction strategies and are consistent with current industry behavior. Costs of CO2 reductions achievable through these policies have been estimated in a range from $10 to $40 per metric ton.[39]
>
> Nuclear generating capacity facilitates CO_2 emission reductions at fossil fuel-fired EGUs by providing carbon-free generation that can

replace generation at those EGUs. Because of their relatively low variable operating costs, nuclear EGUs that are available to operate typically are dispatched before fossil fuel-fired EGUs. Increasing the amount of nuclear capacity relative to the amount that would otherwise be available to operate is therefore a technically viable approach to support reducing CO_2 emissions from affected fossil fuel-fired EGUs.[40]

We have determined that, based on available information regarding the cost and performance of the nuclear fleet, preserving the operation of at-risk nuclear capacity would likely be capable of achieving CO_2 reductions from affected EGUs at a reasonable cost.[41]

In addition to the nuclear generation taken into account in the state goals analysis, any additional new nuclear generating units or uprating of existing nuclear units, relative to a baseline of capacity as of the date of proposal of the emission guidelines, could be a component of state plans.[42]

Demand-Side Energy Efficiency

The fourth building block of the BSER identified by EPA is "cost-effective" demand-side energy efficiency programs.

The purposes of demand-side energy efficiency programs vary; goals include to reduce GHG emissions by reducing fossil-fired generation, help states achieve energy savings goals, save energy and money for consumers and improve electricity reliability. They are typically funded through a small fee or surcharge on customer electricity bills, but can also be funded by other sources, such as from CO2 [allowance auction program] proceeds. [43]

California has been advancing energy efficiency through utility-run demand-side energy efficiency programs for decades and considers energy efficiency "the bedrock upon which climate policies are built." It requires its investor-owned utilities to meet electricity load "through all available energy efficiency and demand reduction resources that are cost-effective, reliable and feasible."[44]

Demand-side energy efficiency programs produce electricity-dependent services with less electricity, and thereby support reduced generation from existing fossil fuel-fired EGUs by reducing the demand for that generation. Reduced generation results in lower CO2 emissions. More than 40 states already have established some form of demand-side energy efficiency polices, and individual states have avoided up to 13 percent of their electricity demand.[45]

New Source Review

The New Source Review[46] (NSR) program was designed to prevent the degradation of air quality from the construction of new facilities or modification of existing facilities which have potentially harmful emissions. Efficiency improvements to power plants that reduce regulated pollutants should not theoretically trigger NSR requirements, unless the improvements result in an increase in emissions (e.g., because the modified plant operates for more hours). EPA has proposed that states be given a primary role with regard to NSR determinations.

> As part of its CAA section 111(d) plan, a state may impose requirements that require an affected EGU to undertake a physical or operational change to improve the unit's efficiency that results in an increase in the unit's dispatch and an increase in the unit's annual emissions. If the emissions increase associated with the unit's changes exceeds the thresholds in the NSR regulations discussed above for one or more regulated NSR pollutants, including the netting analysis, the changes would trigger NSR.
>
> While there may be instances in which an NSR permit would be required, we expect those situations to be few. As previously discussed in this preamble, states have considerable flexibility in selecting varied measures as they develop their plans to meet the goals of the emissions guidelines. One of these flexibilities is the ability of the state to establish the standards of performance in their CAA section 111(d) plans in such a way so that their affected sources, in complying with those standards, in fact would not have emissions increases that trigger NSR. To achieve this, the state would need to conduct an analysis consistent with the NSR regulatory requirements that supports its determination that as long as affected sources comply with the standards of performance in their CAA section 111(d) plan, the source's emissions would not increase in a way that trigger NSR requirements. [47]
>
> The EPA is aware of the potential for "rebound effects" from improvements in heat rates at individual EGUs. In this context, a rebound effect would occur where, because of an improvement in its heat rate, an EGU experiences a reduction in variable operating costs that makes the

EGU more competitive relative to other EGUs and consequently raises the EGU's generation output. The increase in the EGU's CO2 emissions associated with the increase in generation output would offset the reduction in the EGU's CO2 emissions caused by the decrease in its heat rate and rate of CO2 emissions per unit of generation output. The extent of the offset would depend on the extent to which the EGU's generation output increased (as well as the CO2 emission rates of the EGUs whose generation was displaced). The EPA considers the rebound effect to be a potential concern if heat rate improvements were the only approaches being considered for the BSER, but believes that the effect can be addressed by establishing the BSER as a combination of approaches that includes not only heat rate improvements but also approaches that will encourage reductions in electricity demand or increases in generation from lower- or zero-emitting EGUs.[48]

Issues Related to Compliance Strategies

The EPA's proposal for GHG reduction answers questions from the electric power sector with regard to the timeframe, timeline, and choices that would be made available for compliance. The CPP proposal sets out a vision for a greater proportion of electricity coming from natural gas and renewable electric generation, and less from coal-fired power plants, with state-specific goals for carbon emissions reductions proposed for 2030. However, there are still some unresolved issues remaining with potential implementation of the CPP.

Potential Impacts on Retail Electricity Prices

Under the Energy Policy Act of 2005 (P.L. 109-58), security constrained economic dispatch is defined in section 1234 as follows:

> ... the operation of generation facilities to produce energy at the lowest cost to reliably serve consumers, recognizing any operational limits of generation and transmission facilities.

EPA's CPP recognizes that security constrained economic dispatch "assures reliable and affordable electricity."[49] However, some observers say that EPA's CPP essentially proposes an environmental dispatch regime for power plant operation.[50] Under environmental dispatch, the goal is to use "cleaner" power generating units (i.e., which emit the least pollutants) by scheduling these plants to operate first and as much as possible to serve load demands.

Power plants today are generally scheduled to operate (i.e., dispatched) under an economic dispatch regime whereby power generation units are dispatched using generating units with the lowest costs. Thus, under these economic dispatch regimes, the cost of power generation is characterized by a power plant's efficiency or heat rate, its variable costs of generation, its variable costs of environmental compliance, and its start-up costs.[51]

The increased availability of natural gas has resulted in lower prices for wholesale electricity, with a general expectation that wholesale prices will remain relatively low for the next few years.[52] EPA has conceded that this increased demand could push natural gas prices higher.[53]

There is a concern that shifting to an environmental dispatch regime could potentially result in increased electricity prices to consumers, depending on the generation resource mix employed. However, EPA expects only a "modest impact" on retail prices will result.

> As described below in Section X, the results indicate that the proposed state goals (discussed in Section VII) are readily achievable with no adverse impacts on electricity system reliability, and that impacts on retail electricity prices are modest and fall within the range of price variability seen historically in response to changes in factors such as weather and fuel supply.[54]
>
> Retail electricity prices are projected to increase 6 to 7 percent under Option 1 and increase by roughly 4 percent under Option 2, both in 2020 and on an average basis across the contiguous U.S. By 2030 under Option 1, electricity prices are projected to increase by about 3 percent.[55]

Moreover, EPA expects that energy efficiency measures may lead to an actual reduction in the average retail bill for electricity of 9% by 2030.

> Average monthly electricity bills are anticipated to increase by roughly 3 percent in 2020, but decline by approximately 9 percent by 2030. This is a result of the increasing penetration of demand-side programs that more than offset increased prices to end users by their expected savings from reduced electricity use.[56]

Implications for Fuel Diversity

EPA's CPP proposal ostensibly involves natural gas consumption under two of the four legs of the BSER stool. EPA has suggested shifting the dispatch of power generators to lower-emitting sources by increased scheduling of higher efficiency natural gas combined cycle units. Scheduling these plants will result in higher natural gas consumption.[57] EPA has also suggested using more zero-emitting sources by deploying more renewable generation, which in many parts of the United States will require more natural gas consumption to make variable renewable electric generation more firm (i.e., provide power as renewable electric generation ebbs).

As of 2012, electric power generation used 8.5 trillion cubic feet (TCF) of natural gas.[58] EPA's CPP proposal essentially favors a switch to natural gas as the primary fuel used for power generation, and estimates an increase of 1.2 TCF over 2012 consumption in 2020.[59] However, increasing the use of natural gas for power generation raises some concerns, as deliverability and price volatility issues have emerged as recently as this past winter with the demand spikes associated with the Polar Vortex cold weather events.[60] Recovery of costs from the Polar Vortex have proved to be an issue for some utilities,[61] and the performance of demand response programs in periods of extreme weather (e.g., in the winter especially) have come under question.[62]

FERC is working to improve coordination between the electricity and natural gas industries.[63] Electricity generators get their natural gas from major pipelines or local distribution companies, and these deliveries are usually scheduled during nomination cycles.[64] More cost-effective, natural gas storage facilities may be required for electric power production purposes, if greater natural gas use for power generation is expected. However, the regulatory regime (i.e., Regional Transmission Organization markets or traditional regulation) in place will likely have a bearing on what choices are available to natural gas generators with regard to gas storage options or contracting for firm capacity vs. the "just-in-time" manner of natural gas deliveries traditionally available to power generators.

The electric utility industry values diversity in fuel choice options since reliance on one fuel or technology can leave electricity producers vulnerable to price and supply volatility. EPA's proposal expects additional retirements of coal-fired power plants, with some new NGCC capacity likely built to replace retiring coal capacity. Nuclear power plants are also aging, and some plants expected to be in operation in the 2020 to 2030 timeframe could face premature retirement for a variety of reasons ranging from plant age to electricity market or other conditions.

Unless electricity storage capacity is increased or other concepts develop, natural gas will likely be used to smooth the variable output of some renewable electricity technologies. The developing potential for a heavier reliance on natural gas for power generation is a concern for many in the power sector. EPA, for its part, believes that its BSER proposal can help preserve fuel diversity goals.

> Large vertically integrated utilities generally have options within all four building blocks. They tend to have large and, as a general matter, at least somewhat diverse generation fleets. For their higher-emitting units, they have opportunities to use measures that reduce the units' CO2 emission rates, such as heat rate improvements, co-firing, or fuel switching. While this proposal preserves fuel diversity, with over 30 percent of projected 2030 generation coming from coal and over 30 percent from natural gas, even companies that have traditionally depended upon coal to supply the majority of their generation are diversifying their fleets, increasing their opportunities for re-dispatch.[65]

Conversion of Coal to Natural Gas Firing

Switching a coal burning plant to natural gas can be a major undertaking as the boiler, fuel handling, and fuel storage areas would have to be modified or replaced. A major engineering study would have to be undertaken to determine the cost and extent of work to be performed for a specific power plant unit. Coal power plants can have multiple units (i.e., with separate steam boilers), some of varying ages and designs. A power plant location near a major natural gas pipeline would make supplying natural gas to the facility easier, although a local natural gas distribution company could also supply the fuel.

Adding a gas turbine to an existing steam turbine would be one option, but not without challenges or significant modifications based on the age of the existing steam turbine and balance of plant. Moderate increases in plant efficiency are possible from such a modification. A conversion to a combined-cycle configuration could be a major modification, both in terms of work performed and cost. But a considerable increase in power output and efficiency could result from such an upgrade.[66]

Regulatory, Policy, and Reliability Concerns

State-specific compliance plans geared to individual state needs may complicate the coordination necessary for reliability purposes. The individual state compliance plans required by EPA's CPP may have to be submitted to multiple jurisdictions (i.e., state public utility commissions, Regional Transmission Organizations, the North American Electric Reliability Corporation, and FERC) at a number of deliberative levels before a compliance plan can be finalized.[67]

Many fossil-fueled power plants do more than just generate electricity. Many power plants provide ancillary services such as voltage support and frequency regulation to the grid. Additional retirements of coal-fired capacity can impact reserve margins, potentially impacting reliability when needed during weather-related outages or periods of temperature extremes.

Incidents of more extreme weather appear to be occurring, and will need to be planned for when considering the types of future generation which may need to be built to assure electric system reliability.[68] EIA currently expects that a total of 60 GigaWatts of coal capacity will retire by 2020, with 90% of these retirements taking place by 2016 "coinciding with the first year of enforcement for the Mercury and Air Toxics Standards."[69] Much of this capacity scheduled for retirement was dispatched during the recent Polar Vortex, adding concern to how the grid will meet power demands in future weather extremes.[70]

EPA's CPP proposal relies on state-implemented renewable portfolio standards (RPSs) and energy efficiency resource standards going forward. However, many state renewable portfolio standards and goals are scheduled to expire in the 2015 to 2020 timeframe, with more by 2025.[71] And many state RPS policies with mandatory requirements have cost caps to ensure that the targets can be met cost-effectively. Similarly, many state energy efficiency resource standards are expiring by 2020.[72]

The transmission system itself is aging and in need of modernization.[73] The grid is stressed in many regions because the system is being used in a manner for which it was not designed. More transmission capacity will likely be needed to handle potentially more transmission transactions under the EPA proposal. Much of the transmission system was built by individual electric utilities to serve their own power plants. New power plants or increased use of existing NGCC capacity may require upgraded transmission facilities, and potentially new natural gas infrastructure to provide fuel. Increased dependence on renewable generation will likely require new transmission lines, and many of today's transmission projects awaiting regulatory approvals are intended to serve renewable electricity projects.

FERC identified public policy requirements (such as state renewable portfolio standards) as drivers which should be elevated to the level of reliability when it comes to approving new transmission projects in its Order No. 1000, *Transmission Planning and Cost Allocation.*[74] Actual implementation of regional compliance plans will demonstrate whether the regime for transmission planning and cost sharing will achieve FERC's goals.

The focus of the EPA's CPP proposal will arguably fall on coal-fired power plants, with at least two of the four building blocks centering on coal plant efficiency, dispatch, and emissions. The age and condition of coal-fired power plants are key considerations in a decision to upgrade or modify plants. Power plants in Regional Transmission Organization (RTO) regions operate in competitive environments where a power plant's operating and maintenance costs are not guaranteed recovery. The additional costs of plant upgrades may not be cost-effective under RTO electricity market regimes or prices. State implementation plans for EPA's CPP may also result in differing requirements within RTO regions for competitive generators. Capacity markets designed to incentivize the construction of new generation in regions with competitive markets have had mixed results. New power plants will most likely be built in regions of the country with traditional regulation using tools like integrated resource planning, and rules allowing cost recovery from ratepayers for approved investments.[75]

Potential for Varying State Impacts

EPA's state-specific GHG emissions goals vary considerably in magnitude, leading to concerns that some states may have much more to do than others and thus compliance with the CPP could result in "winners and losers." The approach to compliance taken by each state will certainly have a unique cost and economic impact, and these could vary considerably between states. However, EPA asserts the benefits of GHG reduction will far outweigh these costs. The agency says that the state-specific goals reflect each state's unique emissions profile and generation resource mix, and maintains that the flexibility offered by its "building blocks" approach will allow states the opportunity to choose a strategy capable of minimizing compliance costs and economic impacts.

Similarly, we recognize and appreciate that states operate with differing circumstances and policy preferences. For example, states have differing access to specific fuel types, and the variety of types of EGUs operating in different states is broad and significant. States are part of assorted EGU dispatch systems and vary in the amounts of electricity that they import and export. For these reasons, we also recognize and appreciate the value in allowing and promoting multi-state reduction strategies.[76]

CONCLUSION

Moving forward, EPA GHG regulations can provide a basis for the evolution of the U.S. Electric Power Sector. EPA recognizes that the grid and many of its fossil-fueled power plants are aging, and provides input via the CPP as to how a future national system providing cleaner energy choices could be powered. EPA believes the benefits of a cleaner environment from its plan are without question. However, meeting the goals of EPA's proposed plan will effectively require less power generation from coal-fired power plants, or even outright retirements of coal-fired generation. Considering the average age of the coal-fired power plant fleet, more retirements are likely when the costs of efficiency improvements or upgrades are weighed in compliance plans.

EPA is not proposing the adoption of any new technologies, but suggests a framework for transition. Implementing compliance plans will not come without real costs or making hard choices for the states and electric utilities who will have to work together to find an acceptable compromise. The potential implications for reliability, and the ultimate financial costs of the CPP will become clearer as state compliance plans are filed, and implementation plans become known.

End Notes

1 Greenhouse gases, according to EPA, are any gases that absorb infrared radiation in the atmosphere. There are six greenhouse gases addressed by EPA regulatory actions: carbon dioxide (CO2), methane (CH4), nitrous oxide (N2O), and fluorinated gases—sulfur hexafluoride (SF6), hydrofluorocarbons (HFCs), and perfluorocarbons (PFCs). Carbon dioxide is the most prevalent GHG produced by combustion of fossil fuels. See http://www.epa.gov/climatechange/ ghgemissions/gases.html.

[2] See EPA's proposed Clean Power Plan at http://www2.epa.gov/sites/production/files/2014-05/documents/ 20140602proposal-cleanpowerplan.pdf. (CPP)

[3] Environmental Protection Agency, "EPA Proposes First Guidelines to Cut Carbon Pollution from Existing Power Plants/Clean Power Plan is flexible proposal to ensure a healthier environment, spur innovation and strengthen the economy," press release, June 2, 2014, http://www2.epa.gov/carbon-pollution-standards/clean-power-plan-proposedrule.

[4] "Humans tap the huge pool of fossil carbon for energy, and affect the global carbon cycle by transferring fossil carbon—which took millions of years to accumulate underground—into the atmosphere over a relatively short time span. As a result, the atmosphere contains approximately 35% more CO2 today than prior to the beginning of the industrial revolution. As the CO2 concentration grows it increases the degree to which the atmosphere traps incoming radiation from the sun, which further warms the planet." CRS Report RL34059, The Carbon Cycle: Implications for Climate Change and Congress, by Peter Folger.

[5] Executive Office of the President, The President's Climate Action Plan, June 2013, http://www.whitehouse.gov/sites/ default/files/image/president27sclimateactionplan.pdf.

[6] Renewable electricity includes power generation from wind, solar, geothermal, and biomass sources. However, while hydropower is generally considered as renewable, some would argue that hydropower from large dams with impoundments have potentially harmful environmental impacts both upstream and downstream from the dam.

[7] EPA says that to date, 47 states have demand-side efficiency programs, 38 have renewable portfolio standards or goals, and 10 states have market-based greenhouse gas emissions programs.

[8] Comments on the proposed rule must be received on or before October 16, 2014.

[9] Energy Information Administration, "How Much Carbon Dioxide Is Produced When Different Fuels Are Burned?," June 4, 2014, http://www.eia.gov/tools/faqs/faq.cfm?id=73&t=11.

[10] 549 U.S. 497, 529 (2007).

[11] 42 U.S.C. 7401 et seq. (as amended).

[12] U.S. Environmental Protection Agency, Endangerment and Cause or Contribute Findings for Greenhouse Gases under Section 202(a) of the Clean Air Act, November 22, 2013, http://www.epa.gov/climatechange/endangerment/.

[13] EPA, Clean Power Plan Proposed Rule, June 2, 2014, http://www2.epa.gov/carbon-pollution-standards/clean-powerplan-proposed-rule.

[14] U.S. Environmental Protection Agency, "Carbon Pollution Emission Guidelines for Existing Stationary Sources: Electric Utility Generating Units," 79 Federal Register 34829, June 18, 2014. (EMEGU)

[15] Heat rate is the efficiency of conversion from fuel energy input to electrical energy output often expressed in terms of BTU per kiloWatt-hour.

[16] A discussion of potential improvements at coal-fired power plants is presented in CRS Report R43343, Increasing the Efficiency of Existing Coal-Fired Power Plants, by Richard J. Campbell.

[17] EMEGU, Table 8, Proposed State Goals.

[18] One tonne (also known as a metric ton) is a unit of mass equaling 1,000 kilograms.

[19] EPA, EPA Fact Sheet—Cutting Carbon Pollution from Power Plants, June 2, 2014, http://www2.epa.gov/sites/production/files/2014-06/documents/20140602fs-important-numbers-clean-power-plan.pdf.

[20] Edison Electric Institute, Different Regions of the Country Use Different Fuel Mixes to Generate Electricity, 2014, http://www.eei.org/issuesandpolicy/generation/fueldiversity/ Documents/map_fuel_diversity.pdf.

[21] EMEGU, p. 34940.
[22] EMEGU, p. 34863.
[23] Matthew Philips, "EPA Did the Power Industry a Big Favor by Using 2005 Levels," Bloomberg BusinessWeek, June 2, 2014, http://www.businessweek.com/articles/2014-06-02/epa-did-the-power-industry-a-big-favor-by-using-2005- levels.
[24] EMEGU, p. 34867.
[25] EMEGU, p. 34839.
[26] EMEGU, p. 34833.
[27] EMEGU, Table 8.
[28] EPA, Office of Air and Radiation, Goal Computation Technical Support Document, Technical Support Document for the CAA Section 111(d) Emission Guidelines for Existing Power Plants, June 2014, http://www2.epa.gov/sites/ production/files/2014-05/documents/ 20140602tsd-goal-computation.pdf.
[29] EMEGU, p. 34835.
[30] EMEGU, p. 34836.
[31] EMEGU, p. 34838.
[32] EMEGU, p. 34837,
[33] EMEGU, p. 34857.
[34] EMEGU, p. 34862.
[35] EMEGU, p. 34856.
[36] EMEGU, p. 34859.
[37] EMEGU, p. 34857.
[38] "EIA in its most recent Annual Energy Outlook has projected an additional 5.7 GW of capacity reductions to the nuclear fleet.... [EPA views] this 5.7 GW, which comprises an approximately six percent share of nuclear capacity, as a reasonable proxy for the amount of nuclear capacity at risk of retirement." EMEGU, p. 34871.
[39] EMEGU, p. 34858.
[40] EMEGU, p. 34870.
[41] EMEGU, p. 34871.
[42] EMEGU, p. 34923.
[43] EMEGU, p. 34849.
[44] EMEGU, p. 34850.
[45] EMEGU, p. 34858.
[46] NSR was established by Congress as part of the 1977 Clean Air Act Amendments (P.L. 95-95), and is codified in sections 165-169 of the act. NSR requires pre-construction permits and the application of Best Available Control Technology at new major sources of air pollution, and at major modifications of existing major sources.
[47] EMEGU, p. 34928.
[48] EMEGU, p. 34859.
[49] EMEGU, p. 34862.
[50] "Scheduling of power plant operation or intertie access in the order of increasing damage to the environment, with the least environmentally damaging first." See http://www.bpa.gov/ news/pubs/Pages/Definitions---E.aspx.
[51] FERC Staff, Economic Dispatch: Concepts, Practices and Issues, Federal Energy Regulatory Commission, November 13, 2005, http://www.ferc.gov/eventcalendar/ Files/200511101 72953-FERC%20Staff%20Presentation.pdf.
[52] See EIA, Natural Gas Section, at http://www.eia.gov/forecasts/archive/aeo13/ source_natural_ gas_all.cfm#netexporter.

[53] "Under both approaches, the shifting in generation from higher-emitting steam EGUs to lower-emitting NGCC units results in an increase in natural gas production and price. The two-block approach results in a production increase of 19-22 percent and a price increase of 10-11 percent. The four-block approach results in a production increase of 12-14 percent and a price increase of 9-12 percent." EMEGU, p. 34933.

[54] EMEGU, p. 34885.

[55] EMEGU, p. 34935.

[56] EMEGU, p. 34934.

[57] "Given that significant underutilized NGCC exists in various U.S. regions, the possibility of further shifting from coal base load plants to natural gas intermediate capacity exists. A recent study by the Massachusetts Institute of Technology in 2011 noted that the existing U.S. NGCC generation fleet had an average capacity factor of approximately 41%, while its design capacity allowed such plants to operate at 85%. The MIT study looked at a scenario across selected regions of the United States which mimicked the 'full dispatch' of existing natural gas combined cycle plants. The study concluded that under such a scenario (while noting that transmission constraints exist), there is 'sufficient surplus NGCC capacity to displace roughly one-third of U.S. coal generation, reducing CO2 emissions from the power sector by 20%.'" See CRS Report R42950, Prospects for Coal in Electric Power and Industry, by Richard J. Campbell, Peter Folger, and Phillip Brown. (CoalProspects)

[58] EIA Annual Energy Outlook 2014.

[59] EMEGU, p. 34934.

[60] FERC, 2014Winter 2013-2014 Operations and Market Performance in RTOs and ISOs, AD14-8-000, April 1, 2014, http://www.ferc.gov/legal/staff-reports/2014/04-01-14.pdf.

[61] Veronique Bugnion, The Polar Vortex Wreaks Havoc On Utility Bills, Energy Collective, January 31, 2014, http://theenergycollective.com/vbugnion/334481/polar-vortex-wreaks-havoc-utility-bills.

[62] "Not all events requiring commitment of demand resources will occur during the spring, summer, and early fall, when the Limited and Extended Summer Demand products apply. As the Commission's gas-electric coordination investigation has shown, reliability problems can occur during the winter when gas-fired generators may have difficulty with obtaining natural gas or transportation of natural gas. See Communication of Operational Information Between Natural Gas Pipelines and Electric Transmission Operators, Order No. 787, FERC Stats. & Regs. ¶ 31,350 (cross-referenced at 145 FERC ¶ 61,134, at P 8 (2013)) ("short term swings in demand by gas-fired electric generators resulting from redispatch by electric transmission operators may be difficult to manage, particularly during times of coincident peak loads on interstate natural gas pipelines and electric transmission systems, such as during unusual cold weather events when end-use customers may rely on both natural gas and electricity"). See also PJM supplemental answer at 7 (discussing PJM's need for demand response during the polar vortex on January 7 and 8, 2014, and indicating that all of its demand response was Limited Demand Response and therefore could not be required to run)." See 146 FERC ¶ 61,052, footnote 48.

[63] FERC, Natural Gas—Electric Coordination, June 2014, http://www.ferc.gov/industries/electric/indus-act/electriccoord.asp.

[64] "The natural gas industry generally follows the scheduling cycles adopted by the [North American Energy Standards Board (NAESB)], which FERC regulations incorporate by reference. The NAESB standards set a nationwide natural gas operating day (Gas Day), beginning at 9:00 a.m. CCT [Central Clock Time] and ending at 9:00 a.m. CCT the following day. Current regulations provide for a minimum of four standard nomination

cycles over that 24-hour period with a 'Timely Cycle' and 'Evening Cycle' for nominations closing in the prior day and two 'Intra-Day' nominations during the Gas Day." See http://www.vnf.com/2311.

[65] EMEGU, p. 34886.

[66] See Table 1, "Summary of Emission Reduction Scenarios," Congressional Distribution Memorandum CD145, Summary of Studies on Achieving Increased Coal Power Plant Efficiency and Lower Carbon Dioxide Emissions, January 15, 2014, http://rsinquery .loc.gov/crsx/products-nd/14.5.doc.pdf.

[67] See CoalProspects, "Electricity Reliability—State and Market Inputs."

[68] See CRS Report R42696, Weather-Related Power Outages and Electric System Resiliency, by Richard J. Campbell.

[69] Energy Information Administration, "AEO2014 Projects More Coal-Fired Power Plant Retirements by 2016 Than Have Been Scheduled," February 14, 2014, http://www.eia.gov/ todayinenergy/detail.cfm?id=15031.

[70] Matthew L. Wald, "Coal to the Rescue, but Maybe Not Next Winter," The New York Times, March 10, 2014, http://www.nytimes.com/2014/03/11/business/energy-environment/coal-to-the-rescue-this-time.html?_r=0.

[71] Database of State Incentives for Renewables and Efficiency, Renewable Portfolio Standard Policies, March 2013, http://www.dsireusa.org/documents/summarymaps/RPS_map.pdf.

[72] American Council for an Energy-Efficient Economy, State Energy Efficiency Resource Standards, April 2014, http://www.aceee.org/files/pdf/policy-brief/eers-04-2014.pdf.

[73] Dwayne Stradford, The Revitalization, Modernization of the Aging Transmission System, Electric Light & Power, January 1, 2012, http://www.elp.com/articles/2012/01/the-revitalization-modernization-of-the-aging-transmissionsystem.html.

[74] See discussion of Order No. 1000 in CRS Report R41193, Electricity Transmission Cost Allocation, by Richard J. Campbell and Adam Vann.

[75] Generally, an Integrated Resource Plan is a 10- to 20-year look forward at options for meeting future energy demand which is revisited typically every three to five years to help ensure the continued validity of the planning process.

[76] EMEGU, p. 34855.

INDEX

D

G

H

I

J

K

L

M

N

O

P

Q

R

S

T

U

V

W

Z